Cover Image Credit/Copyright Attribution: IIIerlok_Xolms/Shutterstock

Maria Jacob · Cláudia Neves ·
Danica Vukadinović Greetham

Forecasting and Assessing Risk of Individual Electricity Peaks

Maria Jacob
University of Reading
Reading, UK

Danica Vukadinović Greetham
The Open University
Milton Keynes, UK

Cláudia Neves
Department of Mathematics
and Statistics
University of Reading
Reading, UK

https://doi.org/10.1007/978-3-030-28669-9

Preface

At the height of climate crisis, the UK strives to maintain its position at the forefront of the most rapidly decarbonising countries, harnessing efforts to end domestic coal power generation by 2025. The Net-zero initiative is the recent UK contribution to stop global warming. Technology has become ubiquitous and this has prompted a fundamental shift from large-scale centrally controlled energy market to distribution system operators (DSO) taking part in the single-flow energy market. As business and homes shift to less energy- and emissions-intensive activities, sustained by the emergence of affordable renewable energy, opportunities arise for new businesses and new market entries in the energy sector, which has hastened a lot of interest into the prediction of individual electric energy demand. With extreme weather events, inter-connectivity of modern society and information collection and speed with which it propagates, the sector faces mass digital disruption. There will be many challenges going forward, but also opportunities, for coming-together scientific disciplines to devise new solutions to old and new problems.

In this book, that grew out of a co-supervision of a Master dissertation in the forecasting of individual electric demand, we present central concepts of extreme value theory, an area of statistics devoted to studying extreme events. We also list currently the most popular prediction algorithms for short-term forecasting that are normally dispersed across different research literature coming from mathematics, statistics and machine learning. Our main goal is to collect the different concepts needed for peak forecasting of individual electric demand, so they require minimal background knowledge and to present those concepts with a clear view of the assumptions required for their application and their benefits and limitations.

The structure of the book The introductory chapter provides a description of the problem, namely, short-term prediction of electric demand on individual level, and motivation behind it. Our focus on peaks is also explained. The two data-sets that are used in Chap. 5 to illustrate the concepts presented in Chaps. 2–4 are described and basic exploratory analysis of two data-sets is presented.

Chapter 2 starts with linear regression that is a basic ingredient of many different forecasting algorithms. Several methods from time-series data analysis are presented including hugely popular ARIMA models. Recently developed permutation-based methods are included, based on their focus on peaks, and this is up to our knowledge for the first time that those methods have a place of their own in a review of popular methods. We hope that the time will show their usefulness. Support vector machines and artificial neural networks, with examples from both forward feed and recurrent networks, are representing machine learning based methods.

Chapter 3 concerns the probabilistic theory underpinning extreme values of independent and identically distributed observations. In the way it is presented here, this theory relies strongly on the analytic theory of regular variation, following closely the work developed by Laurens de Haan. The content of this chapter will lay the foundations to the stochastic properties and corresponding statistical methodology presented in Chap. 4.

The methodology for inference on extreme values addressed in Chap. 4 has its focus narrowed down, as we go along, to the case of short tails with a finite upper bound to suit the specific application to the Irish smart meter data described in Chap. 1. This class of short-tailed distributions being tackled includes, but is not limited to Beta distributions and alike. We will be working on the max-domain of attraction rather than pretending that the limiting distribution provides an exact fit to the sampled data. This will enable a stretch to those distributions attaining finite boundary despite being attached to the Gumbel domain of attraction, thus endowed with more realistic characteristics than the typified exponential fit. Chapter 4 is drawn to a close with a brief literature review on recent theory for extremes of non-identically distributed random variables.

Finally, in Chap. 5 short-term prediction with the focus on peaks is illustrated comparing methods described in Chap. 2 using a subset of publicly available data from Thames Valley Vision project.

Chapter 1 was written by all three authors. Chapter 2 has Maria Jacob and DVG as authors, and Chap. 3 is authored by CN and Maria Jacob. Chapters 4 and 5 are authored by CN and DVG.

The book is designed for any student or professional who wants to study these topics at a deeper level and assumes a wide range of different technical backgrounds. We hope that the book will be also useful for teaching. While we have attempted to balance mathematical rigour with accessibility to people with different technical backgrounds, the presented techniques are illustrated using the real-life data, and the corresponding code can be found on GitHub.

Reading, UK Maria Jacob
Reading, UK Cláudia Neves
Milton Keynes, UK Danica Vukadinović Greetham

Acknowledgements We would like to thank the UK Engineering and Physical Sciences Research Council (EPSRC) funded Centre for Doctoral Training in Mathematics of Planet Earth at the University of Reading and Imperial College London, for making this work possible (grant no. EP/L016613/1).

DVG would like to thank Scottish and Southern Energy Networks for making the data publicly available, and to her collaborators Dr. Stephen Haben, Dr. Georgios Giasemidis, Dr. Laura Hattam, Dr. Colin Singleton, Dr. Billiejoe (Nathaniel) Charlton, Dr. Maciej Fila and Prof. Peter Grindrod. Mr. Marcus Voss kindly provided and discussed his work on permutation-based measures and algorithms. Knowledge Media Institute at the Open University was friendly and supportive environment for writing parts of this book.

CN is very obliged to the University of Reading for supporting Open Access publication of this book. To Laurens de Haan, she will always be extremely grateful for the ever stimulating conversations and inspirational advice. Many thanks to Chen Zhou, who kindly provided input and shared insight about the scedasis boundary estimation. To Dan Crisan and Jennifer Scott for all the support through the CDT-Mathematics of Planet Earth and often beyond that. CN also takes great pleasure in thanking Dr. Maciej Fila and team at SSE Networks for sharing their insight and understanding on the applied work embedded in Chap. 4.

CN and DVG deepest gratitude go to their dear families, who have witnessed our preoccupation and endured our torments over the course of this project.

Contents

Acronyms

a.s.	Almost sure(ly)
AA	Adjusted Average
ANN	Artificial Neural Networks
ApE	Adjusted *p*-norm Error
AR	Autoregressive
ARIMA	Autoregressive Integrated Moving Average
ARMA	Autoregressive Moving Average
BM	Block Maxima
BRR	Bayesian Ridge Regression
d.f.	Distribution function
DSO	Distribution System Operator(s)
DTW	Dynamic Time Warping
EVI	Extreme value index
EVS	Extreme value statistics
EVT	Extreme value theory
GEV	Generalised Extreme Value
HWT	Holt-Winters-Taylor
i.i.d.	Independent and identically distributed
KDE	Kernel Density Estimation
LCT	Low carbon technologies
LSTM	Long Short-Term Memory
LW	Last Week
MAD	Median Absolute Deviation
MAE	Mean Absolute Error
MAPE	Mean absolute percentage error
MLP	Multi-layer Perceptrons
MLR	Multiple Linear Regression
OLS	Ordinary Least Squares
OSH	Overnight Storage Heating
PDF	Probability density function

PLF	Probabilistic load forecasts
POT	Peaks Over Threshold
QQ	Quantile-Quantile
r.v.	Random variable
RNN	Recurrent Neural Network
SD	Similar Day
SME	Small-to-medium enterprises
STLF	Short-term load forecasts
SVM	Support Vector Machine
SVR	Support Vector Regression
TVV	Thames Valley Vision

Chapter 1
Introduction

Electricity demand or load forecasts inform both industrial and governmental decision making processes, from energy trading and electricity pricing to demand response and infrastructure maintenance. Electric load forecasts allow Distribution System Operators (DSOs) and policy makers to prepare for the short and long term future. For informed decisions to be made, particularly within industries that are highly regulated such as electricity trading, the factors influencing electricity demand need to be understood well. This is only becoming more urgent, as low carbon technologies (LCT) become more prevalent and consumers start to generate electricity for themselves, trade with peers and interact with DSOs.

In order to understand and meet demands effectively, smart grids are being developed in many countries, including the UK, collecting high resolution data and making it more readily accessible. This data allows for better analysis of demand, identification of issues and control of electric networks. Indeed, using high quality forecasts is one of the most common ways to understand demand and with smart meter data, it is possible to know not only how much electricity is required at very high time resolutions, but also how much is required at the substation, feeder and/or household level.

However, this poses new challenges. Mainly, load profiles of individual households and substations are harder to predict than aggregated regional or national load profiles due to their volatile nature. The load profiles at the low voltage level contain irregular peaks, or local maxima, which are smoothed out when averaged across space or time. Once aggregated, the profiles are smoother, and easier to forecast. The work presented in this book outlines the challenge of forecasting energy demand at the individual level and aims to deepen our understanding of how better to forecast peaks which occur irregularly.

Even more uncertainty arises from the drastic changes to the way society has used electricity thus far and will do so in the future. Many communities have moved away

M. Jacob et al., *Forecasting and Assessing Risk of Individual Electricity Peaks*, Mathematics of Planet Earth,
https://doi.org/10.1007/978-3-030-28669-9_1

from gas and coal powered technologies to electrically sourced ones, especially for domestic heating [1]. Moreover, where households and businesses were previously likely to be consumers, government policies incentivising solar energy have lead to an increase in photovoltaic panel installations [2], meaning that the interaction with the electricity grid/ DSO will become increasingly dynamic.

In addition to this, governments are also diversifying the source of electricity generation, i.e. with renewable and non-renewable sources [3, 4] and incentivising the purchase of electric vehicles [5] in a bid to reduce national and global greenhouse emissions. This evolution of societal behaviour as well as governmental and corporate commitments to combat climate change is likely to add more volatility to consumption patterns [6] and thereby increase uncertainty. Most likely the changing climate itself will drive different human behaviours to current ones and introduce yet more unknowns to the problem. Therefore, while the literature on forecasting of electricity load is large and growing, there is a definite need to revisit the topic to address these issues. As demand response, battery control and peer-to-peer energy trading are all very sensitive to peaks at the individual or residential level, particular attention will be given to forecasting the peaks in low-voltage load profiles.

While the change of attention from average load to peak load is not new, a novel approach in terms of electricity load forecasting, is to adapt the techniques from a branch of statistics known as Extreme Value Theory (EVT). We will speak in depth about it in later chapters but we briefly share a sense of its scope and our vision for its application to the electricity demand forecasting literature. We can use the methods from EVT to study the bad-case and the worst-case scenarios, such as blackouts which, though rare, are inevitable and highly disruptive. Not just households [7] but businesses [8] and even governments [9] may be vulnerable to risks from blackouts or power failure. In order to increase resilience and guard against such high impact events, businesses in particular may consider investing in generators or electricity storage devices. However, these technologies are currently expensive and the purchase of these may need to be justified through rigorous cost-benefit analyses. We believe that the techniques presented in this book and that to be developed throughout the course of this project could be used by energy consultants to assess such risks and to determine optimal electricity packages for businesses and individuals.

As one of our primary goals is to study extremes in electricity load profiles and incorporate this into forecasts for better accuracy, we will first consider the forecasting algorithms that are commonly suggested in the literature and how and where these algorithms fail. The latter will be done by (1) considering different error measures (the classic approach in load forecasting) and (2) by studying "heteroscedasticity" in forecasts errors (an EVT approach), which for the moment can be understood as the irregular frequency of large errors or even the inability of the algorithm to predict accurately over time. We will also estimate the upper bound of the demand. We believe that DSOs will be able to use these kinds of techniques to realistically assess what contractual obligations to place upon individual customers and thereby tailor their contracts. They may also prove useful in demand response strategies.

In this book, we will consider two smart meter data sets; the first is from smart meter trials in Ireland and the second is collected as part of the Thames Valley Vision

(TVV) Project in the UK. The Irish smart meter trials is available publicly and so has been used in many journal papers and is a good starting point. However, little information about the households is available. The TVV Project on the other hand is geographically compressed on a relatively small area, allowing weather and other data about the area to be collected. The substation data is available at higher time resolution than the Irish smart meter data and subsequently provides more information with which to build statistical models. Combining both the classic forecasts with the results from EVT, we aim to set benchmarks and describe the extreme behaviour.

While both case studies relate to energy, particularly electricity, the methods presented here are by no means exclusively for this sector; they can be and have been applied more broadly as we will see in later chapters. Thus, the work presented in this book may also serve to illustrate how results from EVT can be adapted to different disciplines. Furthermore, this book may also prove conducive to learning how to visualise and understand large amounts data and checking of underlying assumptions. In order to facilitate adaptations to other applications and generally share knowledge, some of the code used in this work has been made accessible through GitHub[1] so those teaching or attending data science courses may use it to create exercises extending the code, or to run experiments on different data-sets.

1.1 Forecasting and Challenges

Electricity load forecasts can be generated for minutes and hours in advance to years and decades in advance. Forecasts of different lengths assist in different applications, for example forecasts for up to a day ahead are generated for the purpose of demand response or battery control, whereas daily to yearly forecasts may be produced for energy trading, and yearly to decade forecasts allow for grid maintenance and investment planning and informing energy policy (Fig. 1.1).

Most studies in electric load forecasting in the past century have focused on point load forecasting, meaning that at each time point, one value is provided, usually an average. The decision making process in the utility industry relies mostly on expected values (averages) so it is no surprise that these types of forecasts have been the dominant tool in the past. However, market competition and requirements to integrate renewable technology have inspired interest in probabilisitic load forecasts (PLF) particularly for system planning and operations. PLF may use quantiles, intervals and/or density functions [10]. We will review the forecast literature in more detail in Chap. 2, focusing mostly on point/deterministic forecasts. It is worth noting that many of those point-forecast methods can be implemented for quantiles prediction.

It becomes evident from various electric load forecasting reviews presented by Gerwig [11], Alfares and Nazeeruddin [12], Hong and Fan [10], that many algorithms of varying complexity exist in the literature. However, for many reasons they are not always particularly good in predicting peaks [13]. The fundamental idea behind most

[1] https://github.com/dvgreetham/STLF.

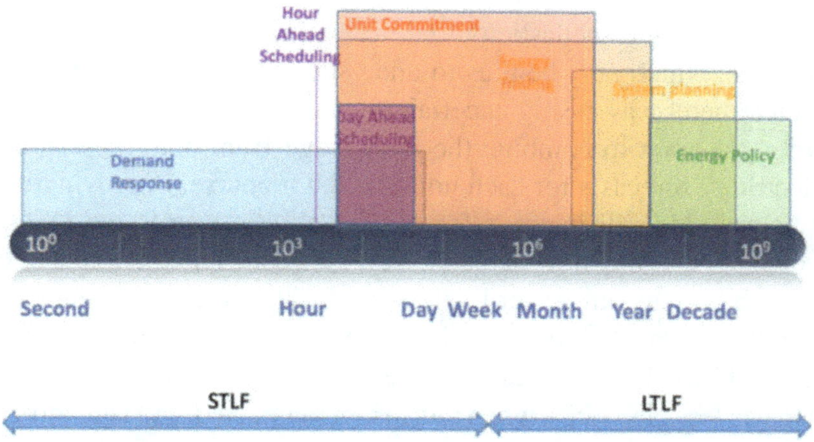

Fig. 1.1 The various classifications for electric load forecasts and their applications. Based on: Hong and Fan [10]. The abbreviations are Short Term Load Forecasting (STLF), and Long Term Load Forecasting (LTLF)

forecasting algorithms is that a future day (or time) is likely to be very much like days (or times) in the past that were similar to it with regard to weather, season, day of the week, etc. Thus, algorithms mostly use averaging or regression techniques to generate forecasts. This brings us back to the first challenge mentioned earlier: such algorithms work well when the demand profiles are smooth, for example due to aggregation at the regional and/or national level, but when the profiles are irregular and volatile, the accuracy of forecasts is reduced. This is usually the case for households or small feeder (sometimes called residential) profiles. In this way, it becomes obvious that we need algorithms that can recreate peaks in the forecasts that are representative of the peaks in the observed profiles.

This brings us to the second challenge: in order to determine which algorithms perform well and which perform better (or worse), we need to establish benchmarks and specify how we measure accuracy. There are many ways of assessing the quality of forecasts, or more strictly many error metrics that may be used. Some conventional error metrics for load forecasts are mean absolute percentage error (MAPE) and mean absolute error (MAE) (see Sect. 2.2.1). These are reasonably simple and transparent and thus quite favourable in the electric load forecasting community. However, as noted by Haben et al. [14], for low-voltage networks, a peaky forecast is more desirable and realistic than a flat one but error metrics such as MAPE unjustly penalise peaky forecasts and can often quantify a flat forecast to be better. This is because the peaky forecast is penalised twice: once for missing the observed peak and again for forecasting it to be where it did not occur, even if only slightly shifted in time. Thus, some other error measures have been devised recently that tackle this issue. We will review these more in Chap. 2.

Both of these challenges can also be approached from an EVT point of view. On the one hand, peaks in the data can be thought of as local extremes. By considering how large the observations can feasibly become in future, we may be able to quantify how likely it is that observations exceed some large threshold. Equally, as discussed before, we can use heteroscedasticity to describe how behaviour deviates from the "typical" in time, which may help us to understand if particular time windows are hard to predict, thereby assessing uncertainty.

Ultimately, we want to combine the knowledge from both these branches and improve electricity forecasts for each household. Of course, improving forecasts of individual households will improve forecasting ability overall, but DSOs are also interested in understanding how demand evolves in time and the limits of consumption. How much is a customer ever likely to use? When are peaks likely to happen? How long will they last? Knowing this at the household level can help DSOs to incentivise flexibility, load spreading or 'peak shaving'. Such initiatives encourage customers to use less electricity when it is in high demand. Load spreading informed only by regional and national load patterns may prove counter productive at the substation level; for example, exclusive night time charging of electric vehicles, as this is when consumption is nationally low, without smart algorithms or natural diversity may make the substations or feeders vulnerable to night time surges, as pointed out in Hattam et al. [15]. Thus, understanding local behaviour is important to both informing policy and providing personalised customer services.

Before we delve into the theory and methods, we familiarise ourselves with Irish smart meter data in Sect. 1.2.1 and with the TVV data in Sect. 1.2.2.

1.2 Data

1.2.1 Irish Smart Meter Data

The first case study uses data obtained from Irish Social Science Data Archive [16]. The Smart Metering Project was launched in Ireland in 2007 with the intention of understanding consumer behaviour with regard to the influence of smart meter technology. To aid this investigation, smart meters were installed in roughly 5000 households. Trials with different interventions were ran for groups of households. The data used in this book are from those households, which were used as controls in the trials. Therefore, they were not taking part in any intervention (above and beyond a smart meter installation). This gives complete measurements for 503 households. We have further subset the data to use only 7 weeks, starting in August 2010, where the weeks are labelled from 16 to 22 (inclusive). No bank holidays or other national holidays were observed in this period. Measurements were taken at half hourly resolution which are labelled from 1 to 48 where 1 is understood to correspond to midnight. Additionally days are also numbered from 593 (16th of August 2010) to 641. From this, the days of the weeks, ranging from 1 to 7 where 1 is Monday and 7 is Sunday, were deduced. Regardless of the number of occupants, each household is considered to be the unit and the terminology of "customer" and "household" are used interchangeably and equivalently throughout.

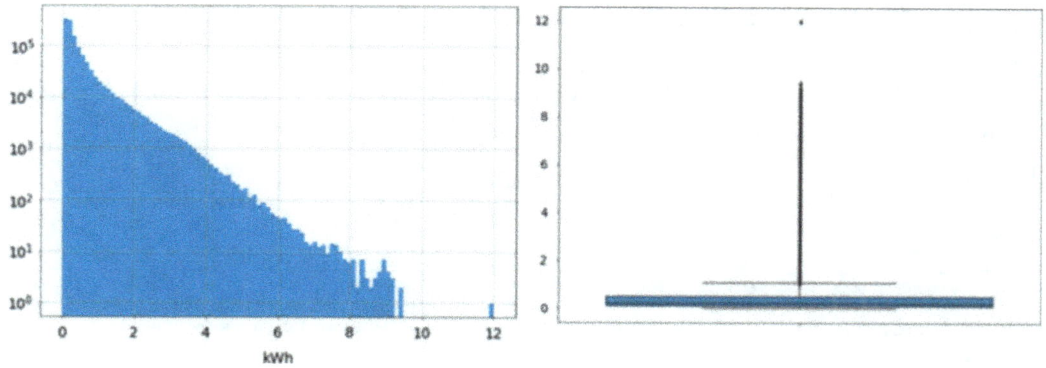

Fig. 1.2 Histogram, logarithmic y scale, and box-plot of half hourly measurements in Irish smart meter data

We now familiarise ourselves with the data at hand. Consider both the histogram and the box plot shown in Fig. 1.2. The 75th percentile for this data is 0.5 kWh meaning that three quarters of the observations are below this value, however some measurements are as high as 12 kWh. Generally, large load values can be attributed to consumers operating a small business from home, having electric heating, multiple large appliances and/or electric vehicles in their home. However, electric vehicle recharging does not seem to be a plausible explanation in this data set as it is a recurring, constant and prolonged activity and such a sustained demand was not observed in any of the profiles. Other large values are roughly between 9 and 10 kWh so we may ask ourselves, what caused such a large surge? Was it a one time thing? How large can that value get within reason? How long can it last? We will address this specific question when we consider "endpoint estimation" in Chap. 4 and for which the theoretical background will be reviewed in Chap. 3.

While Fig. 1.2 tells us about half hourly demand, Fig. 1.3 gives some general profiles. These four plots show the total/cumulative pattern of electricity demand. The top left plot in Fig. 1.3 shows the dip in usage overnight, the increase for breakfast which stabilises during typical working hours with a peak around lunch and rises finally again for dinner, which is when it is at its highest on average. Similarly, the top right plot of Fig. 1.3 shows the total daily consumption for each day in the 7 week period. The plot highlights a recurring pattern which indicates that there are specific days in the week where usage is relatively high and others where it is low. This is further confirmed by the image on the bottom left which tells us that, in total, Fridays tend to have the lowest load, whereas weekends typically have the highest. Finally, the image on the bottom right shows a rise in demand starting in week 18, which is around the beginning of September, aligning with the start of the academic year for all primary and some secondary schools in Ireland. This explains why the jump in data occurs as the weeks preceding are weeks when many families may travel abroad and thus record less electricity demand in their homes.

It is also valuable to see how the top left profile of Fig. 1.3 changes for each day of the week. From Fig. 1.4, it is obvious that there are some differences between weekdays and weekends; the breakfast peak is delayed on weekends but no categor-

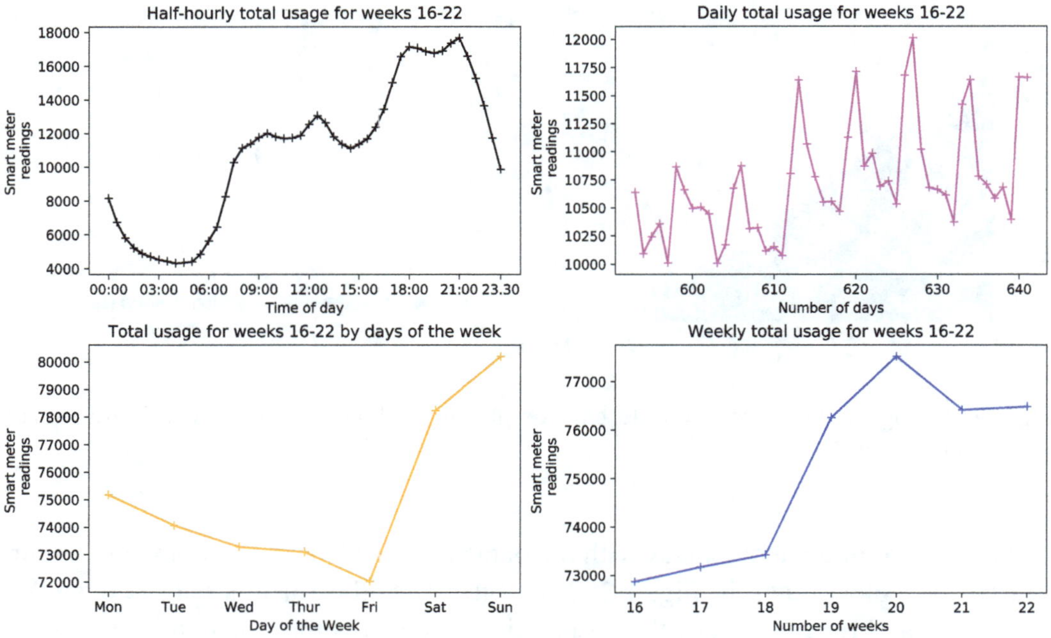

Fig. 1.3 Cumulative demand profiles in kiloWatt hours (kWh) for various time horizons

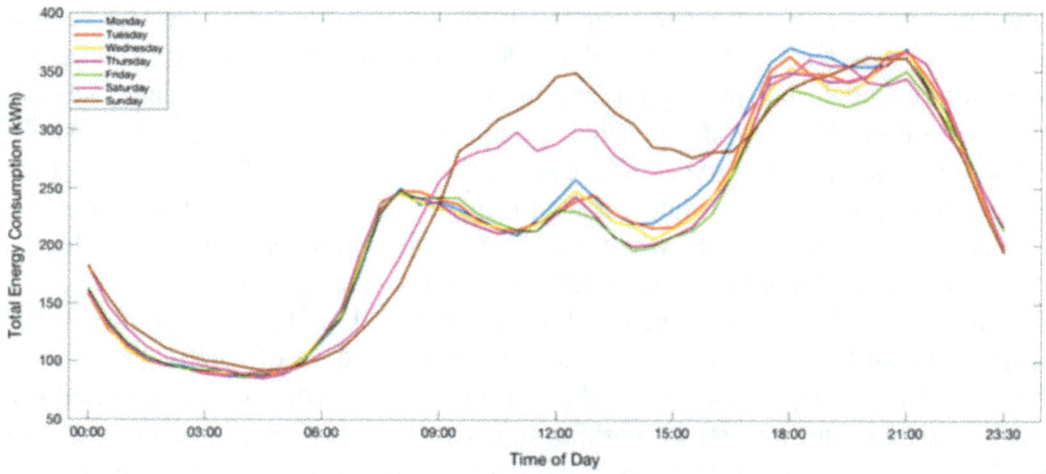

Fig. 1.4 Total load profiles for each day of the week

ical differences are obvious for the evening peaks between weekends and weekdays. Notice that both the top left image of Fig. 1.3 and the weekday profiles in Fig. 1.4 show three peaks: one for breakfast around 8 am, another for lunch around 1 pm and the third in the evening which is sustained for longer. While we are not currently exploring the impact and benefits of clustering, we may use these three identifiers to cluster households by their usage in the future.

Already, we can see the basis for the most forecasting algorithms that we mentioned before. When profiles are averaged, they are smooth and thus overall averaging techniques may work well. Furthermore, if most Sundays record high usage, then it is

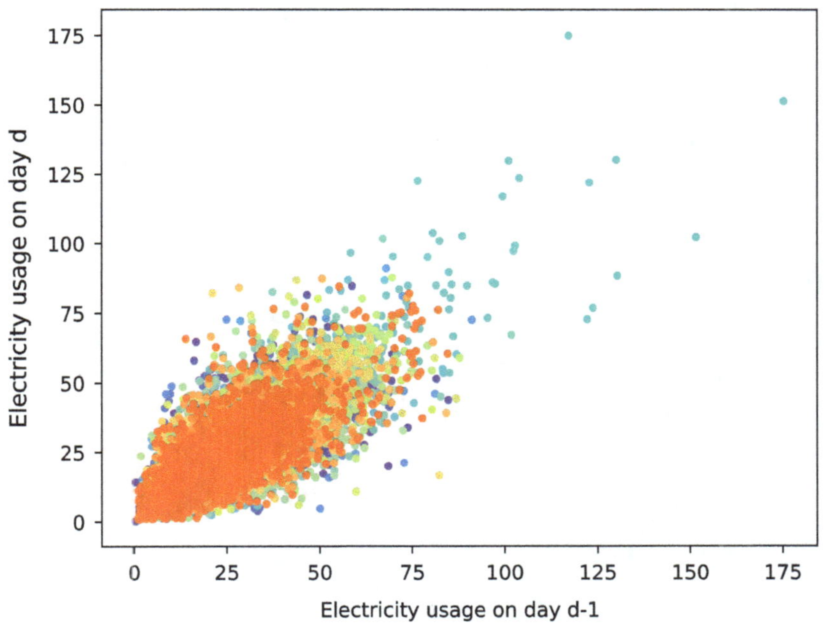

Fig. 1.5 Electric load day d against day $d-1$ in kWh

sensible to use profiles from past Sundays to predict the demand for future Sundays, i.e. to use similar days.

In a similar way, it may be sensible to use similar time windows on corresponding days, that is using past Sunday evenings to predict future Sunday evenings. One way to see if this holds in practice as well as in principle is to consider correlation. Figure 1.5 shows the relationship between the daily demand of each household on day d against the daily demand on day $d-1$. Each marker indicates a different household though it should be noted that there is not a unique colour for each. There seems to be evidence of a somewhat linear trend and some variation which may be resulting from the fact that weekends have not been segregated from weekdays and we are not always comparing similar days. To see how far back this relationship holds, an auto-correlation function (Fig. 1.6) is provided. The auto-correlation function is for the aggregated series given by the arithmetic mean of all customers, $\frac{1}{n}\sum_{i=1}^{n} x_i$, where x_i is the load of the ith household, at each half hour. The dashed line represents the 95% confidence interval. As can be seen, there is some symmetry and while it is not shown here there is also periodicity throughout the data set though with decreasing auto-correlation. This gives us the empirical foundation to use many of the forecasts which rely on periodicity for accuracy.

Finally, and as a prelude to what follows in Chap. 5, one way to see if there are "extreme" households is to consider the daily total demand of each household. This is shown in Fig. 1.7, again with each marker representing different households as before. It is noteworthy that there is one house (coloured in light blue) that consistently appears to be using the most amount of electricity per day. This may be an example of a household where the occupants are operating a small business from home.

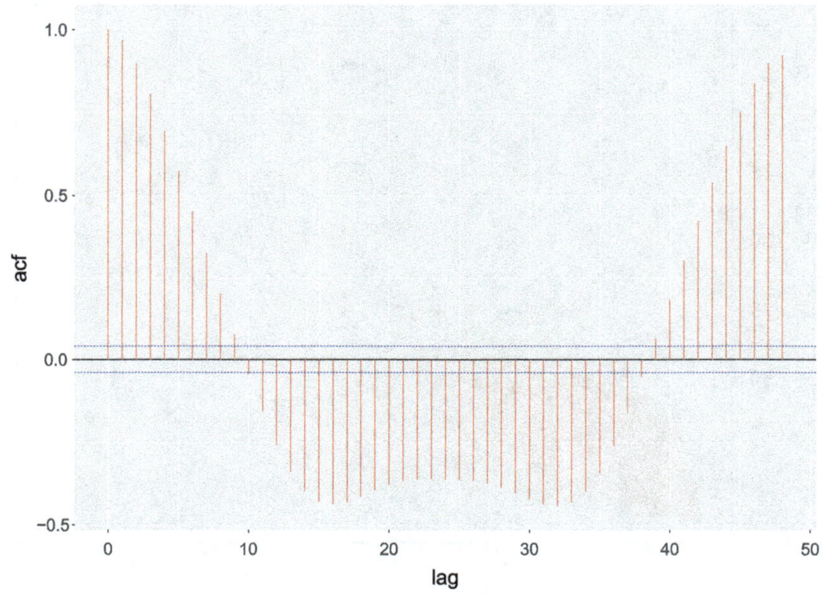

Fig. 1.6 Auto-correlation function for 1 day. Lag is measured in half hour

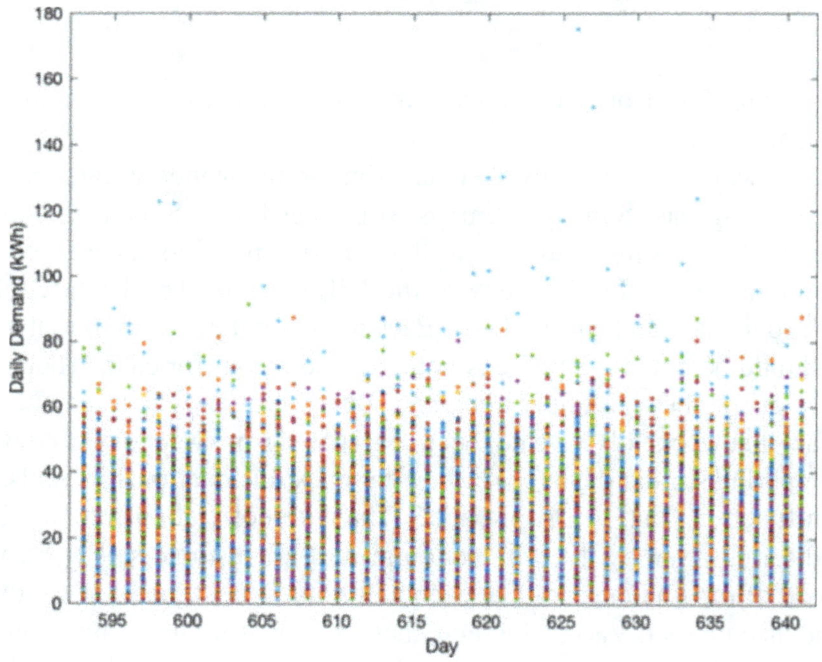

Fig. 1.7 Total daily demand for each household

1.2.2 Thames Valley Vision Data

This second case study uses data that was collected as a part of Scottish and Southern Electricity Network (SSEN) Thames Valley Vision project (TVV),[2] funded by the

[2]http://www.thamesvalleyvision.co.uk/our-project/.

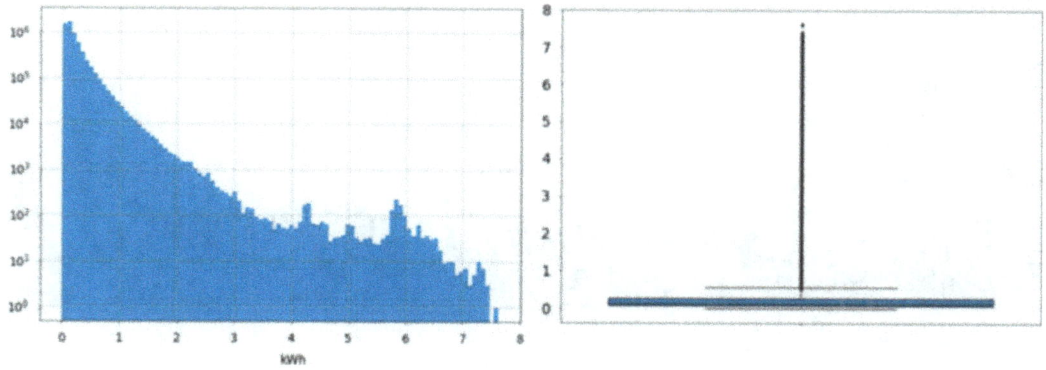

Fig. 1.8 Histogram, logarithmic y scale, and box-plot of half hourly measurements in TVV data

UK gas and electricity regulator Ofgem through the Low Carbon Networks Fund and Network Innovation Competition. The project's overall aim was to monitor and model a typical low voltage network using monitoring in households and substations in order to simulate future realistic demand scenarios. Bracknell, a moderate sized town west of London was chosen as it hosts many large companies and the local network, with its urban and rural parts, is representative of much of Britain's electricity network.

This data set contains profiles for 226 households[3] on half-hourly resolution between 20th March 2014 and 22nd September 2015. The measurements for these households are timestamped and as was done for the Irish smart meter data, information of the day of the week, half hour of the week was deduced. We have also added a period of the week which marks each half hour in a week and ranges from 1, corresponding to 00:15 on Monday, to 336, corresponding to 23:45 on Sunday. We have also subset the data to include only full weeks. Thus, in this section, the analysis is presented for observations taken between 24th March 2014 and 20th September 2015, spanning 546 days which is 78 weeks of data.

We again start by considering the histogram and box plot of all measurements (Fig. 1.8). The largest value in this data set is 7.623 kWh, which is much smaller than our last case study, whereas the 75th percentile is 0.275 kWh. Though the magnitudes of these values are not the same, the general shape of the histogram here is similar to that of the Irish smart meter data; they are both left skewed and large values are relatively few.

The box plot presented in Fig. 1.9 shows the consumption for each household.

Next, we consider the general patterns and trends in the load. We do this by considering the average consumption. Let us start with the top left image of Fig. 1.10. Firstly, it shows that measurements were taken 15 min after and before the hour. The mean profile also appears to be less smooth as expected, than in the case of the Irish smart meter data, as they are less households. Still some fundamental and qualitative similarities persist; on average, electricity demand is low at night. This increases sharply after around 6 am and reaches its peak around 7.45 am. This surge in demand stabilises until a small peak during typical lunch time hours. Again, the

[3] http://data.ukedc.rl.ac.uk/simplebrowse/edc/Electricity/NTVV/EPM.

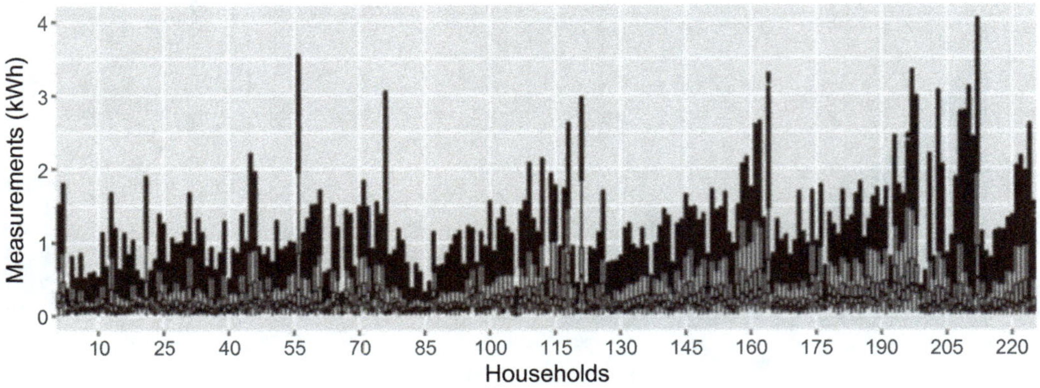

Fig. 1.9 Box-plot of electricity load of each household in the TVV data

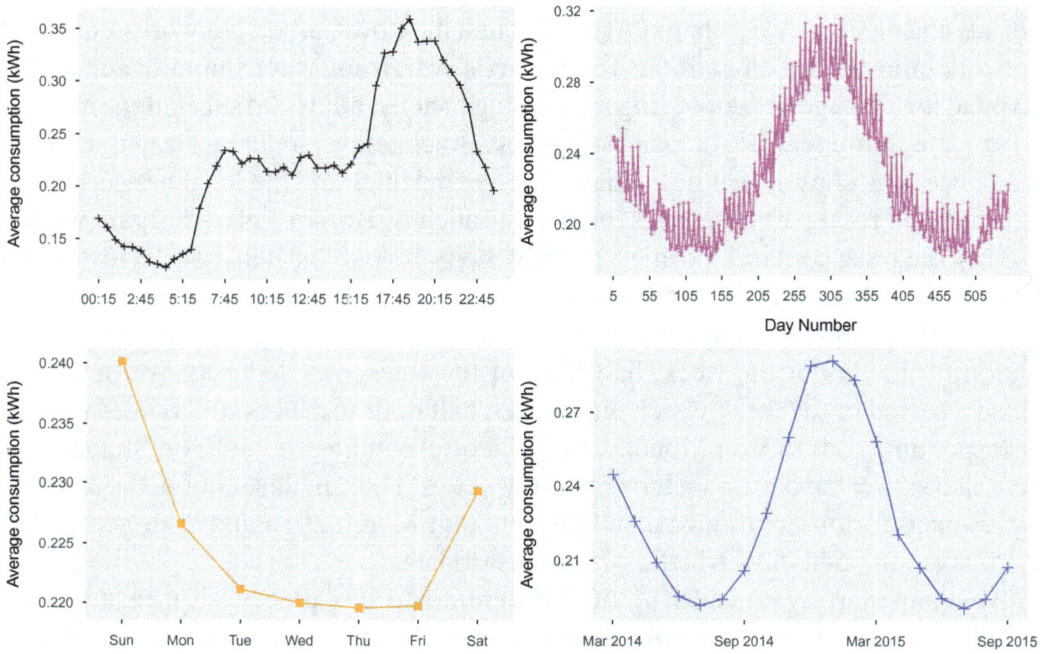

Fig. 1.10 Average demand profiles in (kWh) for various time horizons in the TVV data

evening peak is still the period of highest demand; the peak reaches higher than 0.3 kWh and is sustained for roughly 3 h. Note that if a household has electric vehicle, this will change the demand profile. However, as we discussed before, the presence of electric vehicles will change not just the timing of this high demand but also magnitude and duration.

Of course, these values depend on the time of year and the day of the week as shown in the top right and bottom left plots of Fig. 1.10. The seasonal and annual cycle for daily average demand is obvious from the top right plot. Recall that day 1 corresponds to the 20th of March 2014. Although it would be valuable to have an even longer time series, there are some periods for which two consecutive seasons of data are present. This in general helps in forecasting, because it enables modelling seasonal and annual cycles.

The weekly cycle shown in the bottom left plot of Fig. 1.10 is again in line with what we saw with the Irish smart meter data. On average, the weekends have high electricity consumption with lowest average demands being recorded between Wednesday and Friday. It may be possible that Mondays are relatively high because this plot does not differentiate between Mondays which are weekdays and Mondays which are bank holidays. We will consider this shortly.

Finally, the bottom right plot in Fig. 1.10 reaffirms the seasonal cycle; winter months on average have higher electricity demand than do summer months. This is due to increased lighting, but it is also possible that there are at least some houses in the sample that heat their homes using electricity. Note while this seasonality may be important to model when forecasting aggregated load, it may be less important for forecasting individual load (see e.g. Haben et al. [17], Singh et al. [13], where gas heating is more prominent, like in most parts of the UK (see Department for Business, Energy & Industrial Strategy, UK [18]).

While a day may be classified into the day of the week, we may also classify them by whether it is a holiday or working day and whether it succeeds a holiday or working day. Thus, we now consider how the top left plot of Fig. 1.10 changes depending on such a classification.

The days and times were classified into 4 categories: working day followed by working day (ww), working day followed by a holiday (hw), holiday followed by a working day (wh) and holiday followed by a holiday (hh). All Sundays were classified as "hh" but weekdays can be classified as "hh" for example for Christmas or other bank holidays. Tuesdays to Fridays are mostly qualified as "ww" except when they occur immediate after Easter weekend, Christmas, boxing day, or new year's day, in which case they were classified as "hw". As expected, Saturdays are mostly qualified as "wh" or as "hh" when they succeeded Fridays which were national holidays. The load profiles separated by these day classifications are shown in Fig. 1.11.

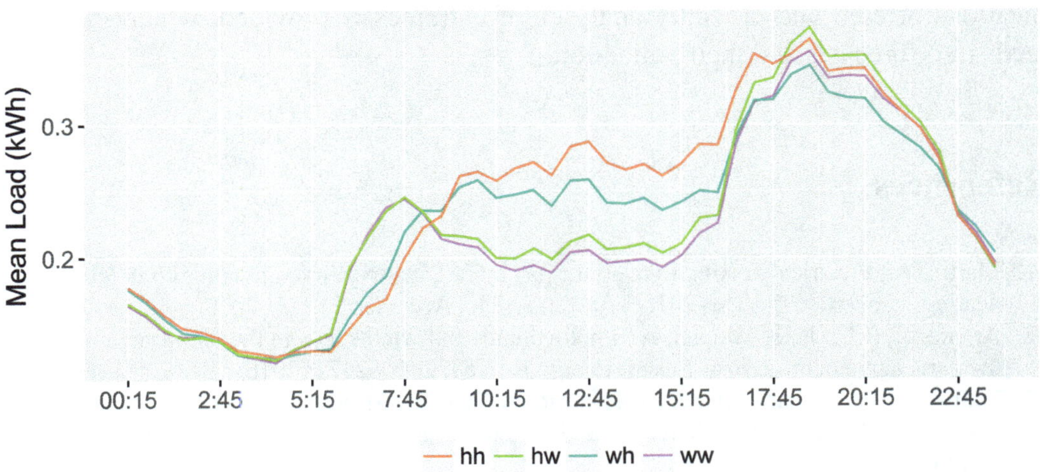

Fig. 1.11 Average demand profiles in (kWh) for each day classification by time of day for TVV

Again, we see qualitatively similar behaviour as the Irish smart meter data; the breakfast peaks occur earlier on working days and at similar times regardless of whether the previous day was a holiday or not. As was the case for the Irish smart meter data, the evening peaks are not distinguishably different between working days and holidays; the main difference is for day time consumption. In general, bank holidays and Sundays have the highest usage, Saturdays and other ordinary non-working days use slightly less but still significantly more than working days. The day time usage on working days is the lowest.

1.3 Outline and Objectives

As was mentioned before, the work that is presented in this book is the first part of a project, which aims to incorporate analyses of extremes into forecasting algorithms to improve the accuracy of forecasts for low-voltage networks, that is substations, feeders and households. Thus, it is an amalgamation of two research areas, which till now have remained relatively separate, in order to inform and affect decision making within the energy industry.

Thus far, we have considered only generally the value of the current line of inference to the utility industry. In what proceeds, we aim to give a thorough review of the literature and provide more specific reasons for why each method is used and discuss its shortcomings. In Chap. 2, we will explore in depth the literature of short term load forecasts (STLF). Within it, we will consider some industry standards, introduce some recent forecasting algorithms, and discuss forecast validation and uncertainty. After that, we will deviate for two chapters into the theory of extremes (Chap. 3) and the statistics of extremes (Chap. 4), both of which form the cornerstones of the work presented in the case studies in Chaps. 4 and 5. Presented forecasting and extremes techniques are illustrated in case studies. Benchmarks for end-point estimators of electric profiles and forecasting algorithms are established, some modifications offered and crucially analyses of extremes is provided, which in return feeds into forecasts and their validation.

References

1. Martin, A.: Electrical heating to overtake gas by 2018. https://www.hvnplus.co.uk/electrical-heating-to-overtake-gas-by-2018/3100730.article. Accessed 25 Feb. 2018
2. Arantegui, R.L., Jäger-Waldau, A.: Photovoltaics and wind status in the european union after the paris agreement. Renew. Sustai. Energy Rev. **81**, 2460–2471 (2018)
3. Evans, S.: Five Charts Show the Historic shifts in UK Energy Last Year (2015). https://www.carbonbrief.org/five-charts-show-the-historic-shifts-in-uk-energy-last-year. Accessed 13 May 2018
4. Department for Business, E., Strategy, I.: Section 5—electricity (2018). https://assets.publishing.service.gov.uk/government/uploads/system/uploads/attachment_data/file/695797/Electricity.pdf. Accessed 13 May 2018

5. Chrisafis, A., Vaughan, A.: France to ban sales of petrol and diesel cars by 2040 (2017). https://www.theguardian.com/business/2017/jul/06/france-ban-petrol-diesel-cars-2040-emmanuel-macron-volvo. Accessed 10 Aug. 2017
6. American Physical Society: Integrating renewable energy on the grid- a report by the aps panel on public affairs (2015). https://www.aps.org/policy/reports/popa-reports/upload/integratingelec.pdf. Accessed 29 May 2019
7. Ghanem, D.A., Mander, S., Gough, C.: I think we need to get a better generator: household resilience to disruption to power supply during storm events. Energy Policy **92**, 171–180 (2016)
8. Campbell, R.J.: Weather-related Power Outages and Electric System Resiliency. CRS report for Congress, Congressional Research Service, Library of Congress (2012)
9. National Research Council: Terrorism and the Electric Power Delivery System. The National Academies Press, Washington, DC (2012)
10. Hong, T., Fan, S.: Probabilistic electric load forecasting: a tutorial review. Int. J. Forecast. **32**(3), 914–938 (2016)
11. Gerwig, C.: Short term load forecasting for residential buildings—an extensive literature review. In: Neves-Silva, R., Jain, L.C., Howlett, R. J. (Eds.) Intelligent Decision Technologies, pp. 181–193, Springer International Publishing, Cham (2015)
12. Alfares, H.K., Nazeeruddin, M.: Electric load forecasting: literature survey and classification of methods. Int. J. Syst. Sci. **33**(1), 23–34 (2002)
13. Singh, R.P., Gao, P.X., Lizotte, D.J.: On hourly home peak load prediction. In: 2012 IEEE Third International Conference on Smart Grid Communications (SmartGridComm), pp. 163–168 (2012)
14. Haben, S., Ward, J., Greetham, D.V., Singleton, C., Grindrod, P.: A new error measure for forecasts of household-level, high resolution electrical energy consumption. Int. J. Forecast. **30**(2), 246–256 (2014)
15. Hattam, L., Vukadinović Greetham, D., Haben, S., Roberts, D.: Electric vehicles and low voltage grid: impact of uncontrolled demand side response. In: 24th International Conference & Exhibition on Electricity Distribution (CIRED), pp. 1073–1076 (2017)
16. Irish Social Science Data Archive (2015). *CER Smart Metering Project*. http://www.ucd.ie/issda/data/commissionforenergyregulationcer/. Accessed 25 July 2017
17. Haben, S., Giasemidis, G., Ziel, F., Arora, S.: Short term load forecasting and the effect of temperature at the low voltage level. Int. J. Forecast. (2018)
18. Department for Business, Energy & Industrial Strategy, UK (2018). National Energy Efficiency Data-Framework (NEED) report: summary of analysis (2018). https://www.gov.uk/government/statistics/national-energy-efficiency-data-framework-need-report-summary-of-analysis-2018

Chapter 2
Short Term Load Forecasting

Electrification of transport and heating, and the integration of low carbon technologies (LCT) is driving the need to know when and how much electricity is being consumed and generated by consumers. It is also important to know what external factors influence individual electricity demand.

Low voltage networks connect the end users through feeders and substations, and thus encompass diverse strata of society. Some feeders may be small with only a handful of households, while others may have over a hundred customers. Some low voltage networks include small-to-medium enterprises (SMES), or hospitals and schools, but others may be entirely residential. Furthermore, local feeders will also likely register usage from lighting in common areas of apartments or flats, street lighting and other street furniture such as traffic lights.

Moreover, the way that different households on the same feeder or substation use electricity may be drastically different. For example, load profiles of residential households will vary significantly depending on the size of their houses, occupancy, socio-demographic characteristics and lifestyle. Profiles will also depend on whether households have solar panels, overnight storage heating (OSH) or electric vehicles [1]. Thus, knowing how and when people use electricity in their homes and communities is a fundamental part of understanding how to effectively generate and distribute electrical energy.

In short term load forecasting, the aim is to estimate the load for the next half hour up to the next two weeks. For aggregated household demand, many different methods are proposed and tested (see e.g. Alfares and Nazeeruddin [2], Taylor and Espasa [3], Hong and Fan [4], etc.). Aggregating the data smooths it, therefore makes it easier to forecast. The individual level demand forecasting is more challenging and comes with higher errors, as shown in Singh et al. [5], Haben et al. [1]. The growth of literature on short term load forecasting at the individual level has started with the wider access to higher resolution data in the last two decades, and is still developing.

M. Jacob et al., *Forecasting and Assessing Risk of Individual Electricity Peaks*, Mathematics of Planet Earth,
https://doi.org/10.1007/978-3-030-28669-9_2

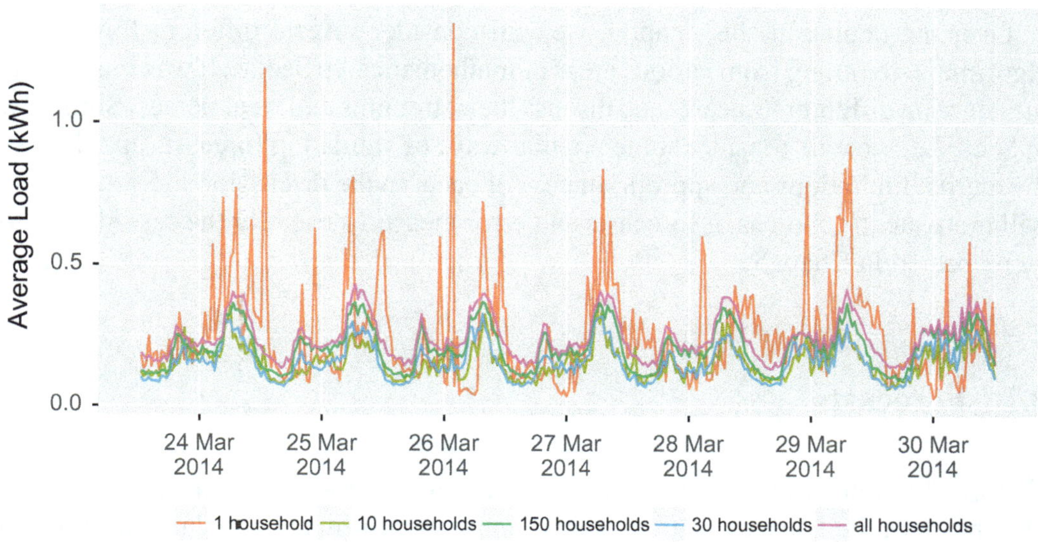

Fig. 2.1 Aggregations of different number of households

So, why is it not enough to look at electricity from an aggregated point of view? Firstly, aggregated load profiles may not reflect individual load profiles, as can be seen from the example in Fig. 2.1. Here, the load has been aggregated for different number of households in one week and the subsequent smoothing is evident. Not only are aggregated load profiles smoother, they also tend to have stronger seasonality and weather dependency than disaggregated load profiles [6]. Demand side response, which encompasses efforts to modify consumption patterns, can be better informed by forecasts which can predict irregular peaks. This is especially true with distributed generation and when both demand and supply become dynamic with LCT integration.

Secondly, as noted by Haben et al. [1], aggregations of individual load profiles do not consider network losses or other loads that are usually not monitored, such as traffic lights and street furniture. Having information of all load allows for better modelling and hence more efficient energy distribution and generation.

Thirdly, considering aggregated load profiles tells us little about individual households or businesses, who may benefit from having tailored energy pricing plans and contracts or need help for informed decision making regarding investment in batteries, photovoltaic and other LCT [7]. The enabling of these kinds of decision making processes is one of the central motivations of the research presented in this book. To do so, we want to consider forecasting methods from both statistics and machine learning literature, specifically the state of the art forecasts within different categories, at the time of writing, and compare them.

In the past, forecasting individual households and feeders was a challenge not just because new forecasting techniques were developing, but also because of the lack of access to a high quality data. The availability of smart meter data alleviates this hindrance and gives new opportunity to address this challenge.

Over the course of this chapter, we will consider several different forecasting algorithms stemming from various areas of mathematics. In Sect. 2.1, we consider the literature on different forecasts and discuss their strengths and weaknesses. Similarly, in Sect. 2.2, we will consider some popular ways of validating forecasts and discuss the merits, limitations and appropriateness of each. In the discussion in Sect. 2.3, we will motivate the choices of forecasts and error measures used for the case studies to be presented in Chap. 5.

2.1 Forecasts

Historically, forecasts have been generated to represent typical behaviour and thus have mostly relied on expectation values. Consequently, many popular algorithms in the literature, and in practice, are point load forecasts using averaging techniques [8] and indeed considering aggregations as mentioned above. Point load forecasts refer to forecasts which give a single, usually mean, value for the future load estimate. Regardless, the need to integrate LCT, market competition and electricity trading have brought about the need for probabilistic load forecasts which may include intervals, quantiles, or densities as noted by Hong and Fan [4] and Haben et al. [1]. In either point load forecasting or probabilistic load forecasting, many approaches exist and increasingly mixed approaches are being used to create hybrid profiles to better represent load with irregular peaks.

The challenge that we are interested in addressing in this chapter is the following: given past load (measured in kWh), we want to create a week-long forecast with the same time resolution as the data, for one or more households. While electricity is consumed continuously, we work with time-stamped, discrete load measurements denoted by y_t where $t = \{1, 2, \ldots, N\}$ denotes time and usually obtained from a smart meter.

In this section, we will review several forecasting algorithms. We will illustrate the analyses presented in this chapter in a case study Sect. 5.1, using the TVV endpoint monitor data described in Sect. 1.2.2.

2.1.1 Linear Regression

Different techniques based on linear regression have been widely used for both short term and long term load forecasting. They are very popular due to the simplicity and good performance in general. Regression is used to estimate the relationship between different factors or predictors and the variable we want to predict. Linear regression assumes that these relationships are linear and tries to find the optimal parameters (or weights) so that the prediction error is minimal. This enables us to easily introduce different kind of variables such as calendar variables, past load and temperature. The basic model for multiple linear regression (MLR) is given by

$$y_t = \boldsymbol{\beta}^T \boldsymbol{x}_t + \epsilon_t, \tag{2.1}$$

where y_t is the dependent variable at time t which is influenced by the p independent variables $\boldsymbol{x}_t = (1, x_{t1}, x_{t2}, \ldots, x_{tp})^T$ and $\boldsymbol{\beta} = (\beta_0, \beta_1, \ldots, \beta_p)^T$ are the corresponding regression parameters. The random error term, ϵ_t, is assumed to be normally distributed with zero mean and constant variance $\sigma^2 > 0$, i.e. $\epsilon_t \mathcal{N}(0, \sigma^2)$. Also, $E(\epsilon_t \epsilon_s) = 0$, for $t \neq s$.

The dependent variable or series is the one we are interested in forecasting, whereas the $\boldsymbol{x}_t$ contains information about the factors influencing the load such as temperature or a special day.

As noted in the tutorial review by Hong and Fan [4], the regressions coefficients or parameters are usually estimated using ordinary least squares using the following formula:

$$\hat{\boldsymbol{\beta}} = \left(\sum_{t=1}^{n} \boldsymbol{x}_t \boldsymbol{x}_t^T \right)^{-1} \sum_{t=1}^{n} \boldsymbol{x}_t \boldsymbol{y}_t \tag{2.2}$$

The least squares estimator for $\boldsymbol{\beta}$ is unbiased, i.e., $\mathbb{E}[\hat{\boldsymbol{\beta}}] = \boldsymbol{\beta}$. We also make the connection that the least squares estimator for $\boldsymbol{\beta}$ is the same as the maximum likelihood estimator for $\boldsymbol{\beta}$ if the errors, ϵ_t, are assumed to be normally distributed.

This simple linear model is then basis for various forecasting methods. We start by listing several examples for aggregated load forecast. For example, Moghram and Rahman [9] explored MLR, amongst others, to obtain 24 h ahead hourly load forecast for a US utility whilst considering dry bulb[1] and dew point temperature[2] as well as wind speed. Two models were calibrated, one for winter and one for summer. The authors divided the day into unequal time zones which corresponded roughly to overnight, breakfast, before lunch, after lunch, evening and night time. It was found that dividing the day in this way resulted in a better fit than not dividing the day at all or dividing it equally. The authors also found significant correlations for temperature and wind speed when using the MLR model.

Charlton and Singleton [10] used MLR to create hourly load forecasts. The regression model considered temperature (up to the power of two), day number, and the multiplication of the two. The created model accounts for the short term effects of temperature on energy use, long term trends in energy use and the interaction between the two. Further refinements were introduced by incorporating smoothed temperature from different weather stations, removing outliers and treating national holidays such as Christmas as being different to regular holidays. Each addition resulted in reduction in errors.

---

[1] Dry bulb temperature is the temperature as measured when the thermometer is exposed to air but not to sunlight or moisture. It is associated with air temperature that is most often reported.

[2] Dew point temperature is the temperature that would be measured if relative humidity is 100% and all other variables are unchanged. Dew point temperature is always lower than the dry bulb temperature.

In a similar vein to the papers above, Alfares and Nazeeruddin [2] consider nine different forecasting algorithms in a bid to update the review on forecasting methods and noted the MLR approach to be one of the earliest. The aim was to forecast the power load at the Nova Scotia Power Corporation and thus pertained to aggregated load. They found the machine learning algorithms to be better overall. The vast literature surveyed in Alfares and Nazeeruddin [2], Hong and Fan [4] and many other reviews, show linear regression to be popular and reasonably competitive despite its simplicity.

While the most common use of regression is to estimate the mean value of the dependent variable, when the independent variables are fixed, it can be also used to estimate quantiles [1, 11].

The simple seasonal quantile regression model used in Haben and Giasemidis [11] was updated in Haben et al. [1] and applied to hourly load of feeders. Treating each half-hour and week day as separate time-series, the median quantile is estimated using the day of trial, three seasons (with sin and cos to model periodicity), a linear trend and then temperature is added using a cubic polynomial.

To find optimal coefficients for linear regression, one usually relies on ordinary least squares estimator. Depending on the structure of a problem, this can result in an ill-posed problem. Ridge regression is a commonly used method of regularisation of ill-posed problems in statistics. Suppose we wish to find an $\boldsymbol{x}$ such that $A\boldsymbol{x} = \boldsymbol{b}$, where A is a matrix and $\boldsymbol{x}$ and $\boldsymbol{b}$ are vectors. Then, the ordinary least squares estimation solution would be obtained by a minimisation of $||A\boldsymbol{x} - \boldsymbol{b}||_2$. However for an ill-posed problem, this solution may be over-fitted or under-fitted. To give preference to a solution with desirable properties, the regularisation term $||\Gamma\boldsymbol{x}||_2$ is added so that the minimisation is of $||A\boldsymbol{x} - \boldsymbol{b}||_2 + ||\Gamma\boldsymbol{x}||_2$. This gives the solution $\hat{\boldsymbol{x}} = \left(A^T A + \Gamma^T \Gamma\right)^{-1} A^T \boldsymbol{b}$.[3]

2.1.2 Time Series Based Algorithms

The key assumptions in classical MLR techniques is that the dependent variable, y_t, is influenced by independent predictor variables $\boldsymbol{x}_t$ and that the error terms are independent, normally distributed with mean zero and constant variance. However, these assumptions, particularly of independence, may not hold, especially when measurements of the same variable are made in time, say owing to periodic cycles in the natural world such as seasons or in our society such as weekly employment cycles or annual fiscal cycle. As such, ordinary least squares regression may not be appropriate to forecast time series. Since individual smart meter data may be treated as time series, we may borrow from the vast body of work that statistical models provide, which allow us to exploit some of the internal structure in the data. In this section, we will review the following time series methods: autoregressive (AR)

[3] In the Bayesian interpretation, simplistically this regularised solution is the most probable solution given the data and the prior distribution for $\boldsymbol{x}$ according to Bayes' Theorem.

models (and their extensions), exponential smoothing models and kernel density estimation (KDE) algorithms.

2.1.2.1 Autoregressive Models

Time series that stem from human behaviour usually have some temporal dependence based on our circadian rhythm. If past observations are very good indicators of future observations, the dependencies may render linear regressions techniques an inappropriate forecasting tool. In such cases, we may create forecasts based on autoregressive (AR) models. In an AR model of order p, denoted by AR(p), the load at time t, is a sum of a linear combination of the load at p previous times and a stochastic error term:

$$y_t = a + \sum_{i=1}^{p} \phi_i y_{t-i} + \epsilon_t, \tag{2.3}$$

with a is a constant, ϕ_i are AR parameters to be estimated, p is the number of historical measurements used in the estimation and ϵ denotes the error term which is typically assumed to be independent with mean 0 and constant variance, σ^2. In a way, we can see some similarity between the MLR model and the AR model; in the MLR, load is dependent on external variables but in the AR model, load is a linear combination of previous values of load.

An example of using an AR model to estimate feeders' load is given in Haben et al. [1]. Here, the model is applied to residuals of load, $r_t = y_t - \mu_t$, where μ_t is an expected value of weekly load. The most obvious advantage of using the residuals is that we can define r_t in a such way that it can be assumed to be stationary. In addition, μ_t models typical weekly behaviour and thus changing the definition of μ_t allows the modeller to introduce seasonality or trends quite naturally and in various different ways, as opposed to the using load itself. In Haben et al. [1], the AR parameters were found using the Burg method.[4] Seasonality can be introduced by including it in the mean profile, μ_t.

Other examples of AR models and their modifications include Moghram and Rahman [9], Alfares and Nazeeruddin [2], Weron [12] and Taylor and McSharry [13], but most of these are studies with aggregated load profiles.

Since we expect that past load is quite informative in understanding future load, we expect that AR models will be quite competitive forecasts, especially when built to include trends and seasonality.

[4]The Burg method minimises least square errors in Eq. (2.3) and similar equation which replaces r_{t-i} with r_{t+i}.

2.1.2.2 Seasonal Autoregressive Integrated Moving Average—SARIMA Models

From their first appearance in the seminal Box & Jenkins book in 1970 (for the most recent edition see Box et al. [14]), autoregressive integrated moving average (ARIMA) time series models are widely used for analysis and forecasting in a wide range of applications. The time series y_t typically consists of trend, seasonal and irregular components. Instead of modelling each of the components separately, trend and seasonal are removed by differencing the data. The resulting time series is then treated as stationary (i.e. means, variances and other basic statistics remain unchanged over time). As we have seen in the previous section, AR models assume that the predicted value is a linear combination of most recent previous values plus a random noise term. Thus,

$$y_t = a + \sum_{i=1}^{p} \phi_i y_{t-i} + \epsilon_t,$$

where a is a constant, ϕ are weights, p is a number of historical values considered and $\epsilon_t \sim \mathcal{N}(0, \sigma^2)$. The moving average model (MA) assumes the predicted value to be the linear combination of the previous errors plus the expected value and a random noise term, giving

$$y_t = \mu + \sum_{i=1}^{q} \theta_i \epsilon_{t-i} + \epsilon_t,$$

where μ is the expected value, θ are weights, q is the number of historical values considered, and $\epsilon_t \sim \mathcal{N}(0, \sigma^2)$. The main parameters of the model are p, d and q, where p is the number of previous values used in the auto-regressive part, d is the number of times we need to difference the data in order to be able to assume that is stationary, and q is the number of previous values used in the moving average part. When strong seasonality is observed in the data, a seasonal part, modelling repetitive seasonal behaviour, can be added to this model in a similar fashion, containing its own set of parameters P, D, Q. A SARMA model, seasonal autoregressive moving average model for 24 aggregated energy profiles is explored in Singh et al. [5] based on 6 s resolution data over a period of one year. Routine energy use is modelled with AR part and stochastic activities with MA part. A daily periodic pattern is captured within a seasonal model. The optimal parameters were determined as $p = 5$ and $q = 30$. The least error square minimisation was used, where the results with different parameter values were compared and the ones that minimised the error were picked up. Interestingly, SARMA not only outperformed other methods (support vector, least square support vector regression and artificial neural network with one hidden layer of ten nodes) regarding mean load prediction, but also regarding peak load prediction, resulting in smaller errors for peaks.

(S)ARMA and (S)ARIMA models can be extended using exogenous variables such as temperature, wind chill, special day and similar inputs. These are called

(S)ARIMAX or (S)ARMAX models, for example Singh et al. [5] gives the following ARMAX model

$$y_t = a + \sum_{i=1}^{p} \phi_i y_{t-i} + \epsilon_t + \mu + \sum_{i=1}^{q} \theta_i \epsilon_{t-i} + \sum_{i=1}^{r} \beta_i T_{t-i},$$

where βs are further parameters that represent the exogenous variables T_t, for instance the outdoor temperature at time t. Two simple algorithms **Last Week (LW)** and **Similar Day (SD)** can be seen as trivial (degenerate) examples of AR models with no error terms, and we will used them as benchmarks, when comparing different forecasting algorithms in Chap. 5. The Last Week (LW) forecast is a very simple forecast using the last week same half-hour load to predict the current one. Therefore, it can be seen as an AR model where $p = 1, d = 0, a = 0, \phi_1 = 1, \epsilon_t \equiv 0$.

The **Similar Day (SD)** forecast instead uses the average of n last weeks, same half-hour loads to predict the current one. Therefore, it can be seen as an AR model where $p = n, d = 0, a = 0, \phi_1, \ldots \phi_n = \frac{1}{n}, \epsilon_t \equiv 0$.

2.1.2.3 Exponential Smoothing Models

The simplest exponential smoothing model puts exponentially decreasing weights on past observations.

Suppose we have observations of the load starting from time $t = 1$, then the single/simple exponential smoothing model is given by

$$S_t = \alpha y_t + (1 - \alpha) S_{t-1}, \tag{2.4}$$

where $\alpha \in (0, 1)$, S_t is the output of the model at t and the estimate for the load at time $t + 1$. Since the future estimates of the load depend on past observations and estimates, it is necessary to specify S_1. One choice for $S_1 = y_1$, but this puts potentially unreasonable weight on early forecasts. One may set S_1 to be the mean of the first few values instead, to circumvent this issue. Regardless, the smaller the value of α, the more sensitive the forecast is to the initialisation.

In the single exponentially smoothed model, as α tends to zero, the forecast tends to be no better than the initial value. On the other hand, as α tends to 1, the forecast is no better than the most recent observation. For $\alpha = 1$, it becomes the LW forecast given in the previous section. The choice for α may be made by the forecaster, say from previous experience and expertise or it may chosen by minimising error functions such as a mean square error.

When the data contains a trend, a double exponential smoothing model is more suitable. This is done by having two exponential smoothing equations: the first on the overall data (2.5) and the second on the trend (2.6):

$$S_t = \alpha y_t + (1-\alpha)(S_{t-1} + b_{t-1}), \tag{2.5}$$
$$b_t = \beta(S_t - S_{t-1}) + (1-\beta)b_{t-1}, \tag{2.6}$$

where b_t is the smoothed estimate of the trend and all else remains the same. We now have a second smoothing parameter β that must also be estimated. Given the model in (2.5) and (2.6), the forecast for load at time $t+m$ is given by $y_{t+m} = S_t + mb_t$.

Of course, we know that electricity load profiles have daily, weekly and annual cycles. Taylor [15] considered the triple exponential smoothing model, also known as the Holt-Winters exponential smoothing model, to address the situation where there is not only trend, but intraweek and intraday seasonality. The annual cycle is ignored as it is not likely to be of importance for forecasts of up to a day ahead. Taylor [15] further improved this algorithm by adding an AR(1) model to account for correlated errors. This was found to improve forecast as when the triple exponential model with two multiplicative seasonality was used, the one step ahead errors still had large auto-correlations suggesting that the forecasts were not optimal. To compensate, the AR term was added to the model.

Arora and Taylor [16] and Haben et al. [1] used a similar model, though without trend, to forecast short term load forecast of individual feeders with additive intraday and intraweek seasonality. Haben et al. [1] found that the so called **Holt-Winters-Taylor (HWT)** triple exponential smoothing method that was first presented in Taylor [17] was one of their best performing algorithms regardless of whether the temperature was included or omitted.

A model is given by the following set of equations:

$$\begin{aligned} y_t &= S_{t-1} + d_{t-s_1} + w_{t-s_2} + \phi e_{t-1} + \epsilon_t, \\ e_t &= y_t - (S_{t-1} + d_{t-s_1} + w_{t-s_2}), \\ S_t &= S_{t-1} + \lambda e_t, \\ d_t &= d_{t-s_1} + \delta e_t, \\ w_t &= w_{t-s_2} + \omega e_t, \end{aligned} \tag{2.7}$$

where y_t is the load, S_t is the exponential smoothed variable often referred to as the level, w_t is the weekly seasonal index, d_t is the daily seasonal index, $s_2 = 168$, $s_1 = 24$ (as there are 336 half-hours in a week and 48 in a day), e_t is the one step ahead forecast error. The parameters λ, δ and ω are the smoothing parameters. This model has no trend, but it has intraweek and intraday seasonality. The above mention literature suggests that when an exponential model is applied, the one-step ahead errors have strong correlations that can be better modelled with an AR(1) model, which in (2.7) is done through the ϕ term. The k-step ahead forecast is then given by $S_t + w_{t-s_2+k} + \phi^k e_t$ from the forecast starting point t.

2.1.2.4 Kernel Density Estimation Methods

Next, we briefly consider the kernel density estimation (KDE), that is quite popular technique in time-series predictions, and has been used for load prediction frequently. The major advantage of KDE based forecasts is that they allow the estimation of the entire probability distribution. Thus, coming up with probabilistic load forecasts is straight forward and results are easy to interpret. Moreover, a point load forecast can be easily constructed, for example by taking the median. This flexibility and ease of interpretation make kernel density forecasts useful for decision making regarding energy trading and distribution or even demand side response. However, calculating entire probability density functions and tuning parameters can be computationally expensive as we will discuss shortly.

We divide the KDE methods into two broad categories, conditional and unconditional. In the first instance, the unconditional density is estimated using historical observations of the variable to be forecasted. In the second case, the density is conditioned on one or more external variables such as time of day or temperature. The simplest way to estimate the unconditional density using KDE is given in (2.8):

$$\hat{f}(l) = \sum_{i=1}^{t} K_{h_y}(y_i - l), \tag{2.8}$$

where $\{y_1, \ldots, L_t\}$ denotes historical load observations, $K_h(\cdot) = K(\cdot/h)/h$ denotes the kernel function, $h_L > 0$ is the bandwidth and $\hat{f}(y)$ is the local density estimate at point y which takes any value that the load can take. If instead, we want to estimate the conditional density, then:

$$\hat{f}(l|x) = \frac{\sum\limits_{i=1}^{t} K_{h_x}(X_i - x) K_{h_L}(y_i - l)}{\sum\limits_{i=1}^{y} K_{h_x}(X_i - x)}, \tag{2.9}$$

where $h_L > 0$ and $h_x > 0$ are bandwidths. KDE methods have been used for energy forecasting particularly in wind power forecasting but more recently Arora and Taylor [18] and Haben et al. [1] used both conditional and unconditional KDE methods to forecast load of individual low voltage load profiles. Arora and Taylor [18] found one of the best forecasts to be using KDE with intraday cycles as well as a smoothing parameter. However, Haben et al. [1] chose to exclude the smoothing parameter as its inclusion costs significant computational efforts. In general, the conditional KDE methods have higher computational cost. This is because the optimisation of the bandwidth is a nonlinear which is computationally expensive and the more variables on which the density is estimated, the more bandwidths must be estimated.

In the above discussion, we have omitted some details and challenges. Firstly, how are bandwidths estimated? One common method is to minimise the difference between the one step ahead forecast and the corresponding load. Secondly, both

Arora and Taylor [18] and Haben et al. [1] normalise load to be between 0 and 1. This has the advantage that forecast accuracy can be more easily discussed across different feeders and this also accelerates the optimisation problem. However, (2.8) applies when l can take any value and adjustments when the support of the density is finite. Arora and Taylor [18] adjust the bandwidth near the boundary whereas Haben et al. [1] do not explicitly discuss the correction undertaken. The choice of kernels in the estimation may also have some impact. The Gaussian kernel[5] was used in both of the papers discussed above but others may be used, for example Epanechnikov[6] or biweight[7] kernels.

2.1.3 Permutation Based Algorithms

Though the methods discussed in the above section are widely used forecasting tools, their performances on different individual smart meter data-sets vary. Some of the mentioned algorithms have smoothing properties and thus, they may be unsuitable when focusing on individual peak prediction. We now list several permutation-based algorithms that are all based on the idea that people do same things repeatedly, but in slightly different time periods. This is of relevance for modelling demand peaks.

2.1.3.1 Adjusted Average Forecast

One of the simple forecasts we mentioned before at the end of Sect. 2.1.2.1, Similar day (SD) forecast averages over the several previous values of load. For example, to predict a load on Thursday 6.30 pm, it will use the mean of several previous Thursdays 6.30 pm loads. But what happens if one of those Thursdays, a particular household is a bit early (or late) with their dinner? Their peak will move half an hour or hour earlier (or later). Averaging over all values will smooth the real peak, and the mistake will be penalised twice, once for predicting the peak and once for missing earlier (later) one. Haben et al. [19] introduced a new forecasting algorithm which iteratively updates a base forecast based on average of previous values (as the SD forecasting), but allows permutations within a specified time frame. We shall refer to it as the **Adjusted Average (AA)** forecast. The algorithm is given as follows:

(i) For each day of the week, suppose daily profiles $\boldsymbol{G}^{(k)}$ are available for past N weeks, where $k = 1, \ldots, N$. By convention, $\boldsymbol{G}^{(1)}$ is the most recent week.
(ii) A base profile, $\boldsymbol{F}^{(1)}$, is created whose components are defined by the median of corresponding past load.

---

[5] $K(x) = \frac{1}{\sqrt{2\pi}} e^{-\frac{1}{2}x^2}$.

[6] $K(u) = \frac{3}{4}(1 - u^2)$ for $|u| \leq 1$ and $K(u) = 0$ otherwise.

[7] $K(u) = \frac{15}{16}(1 - u^2)^2$ for $|u| \leq 1$ and $K(u) = 0$ otherwise.

(iii) This baseline is updated iteratively in the following way. Suppose, at iteration k, we have $\boldsymbol{F}^{(k)}$ for $1 \leq k \leq N-1$, then $\boldsymbol{F}^{(k+1)}$ is obtained by setting $\boldsymbol{F}^{(k+1)} = \frac{1}{k+1}\left(\hat{\boldsymbol{G}}^{(k)} + k\boldsymbol{F}^{(k)}\right)$, where $\hat{\boldsymbol{G}}^{(k)} = \hat{P}\boldsymbol{G}^{(k)}$ with $\hat{P} \in \mathcal{P}$ being a permutation matrix s.t. $||\hat{P}\boldsymbol{G}^{(k)} - \boldsymbol{F}^{(k)}||_4 = \min_{P\in\mathcal{P}} ||P\boldsymbol{G}^{(k)} - \boldsymbol{F}^{(k)}||_4$. $\mathcal{P}$ is the set of restricted permutations i.e, for a chosen time window, ω, the load at half hour i can be associated to the load at half hour j if $|i-j| \leq \omega$.

(iv) The final forecast is then given by $\boldsymbol{F}^{(N)} = \frac{1}{N+1}\left(\sum_{k=1}^{N}\hat{\boldsymbol{G}}^{(k)} + \boldsymbol{F}^{(1)}\right)$.

In this way, the algorithm can permute values in some of historical profiles in order to find the smallest error between observed and predicted time series. This displacement in time can be reduced to an optimisation problem in bipartite graphs, **the minimum weight perfect matching in bipartite graphs**, [20], that can be solved in polynomial time.

A graph $G = (V, E)$ is bipartite if its vertices can be split into two classes, so that all edges are in between different classes. Two bipartite classes are given by observations y_t and forecasts f_t, respectively. Errors between observations and forecasts are used as weights on the edges between the two classes. Instead of focusing only at errors $e_t = y_t - f_t$ (i.e. solely considering the edges between y_t and f_t), differences between

$$y_t - f_{t-1}, y_t - f_{t+1}, y_t - f_{t-2}, y_t - f_{t+2}, \ldots, y_t - f_{t-\omega}, y_t - f_{t+\omega},$$

are also taken into account, for some plausible time-window ω. It seems reasonable not to allow, for instance, to swap morning and evening peaks, so ω should be kept small.

These differences are added as weights and some very large number is assigned as the weight for all the other possible edges between two classes, in order to stop permutations of points far away in time. Now, the perfect matching that minimises the sum of all weights, therefore allowing possibility of slightly early or late forecasted peaks to be matched to the observations without the double penalty is found. The minimum weighted perfect matching is solvable in polynomial time using the Hungarian algorithm Munkres [21], with a time complexity of $O(n(m + n\log n))$ for graphs with n nodes (usually equal to 2×48 for half-hourly daily time series and m edges ($\approx 2 \times n \times \omega$). It is important to notice that although each half-hour is considered separately for prediction, the whole daily time series is taken into account, as permutations will affect adjacent half-hours, so they need to be treated simultaneously.

2.1.3.2 Permutation Merge

Based on a similar idea, Permutation Merge (PM) algorithm presented in Charlton et al. [22] uses a faster optimisation—the minimisation of p-adjusted error (see

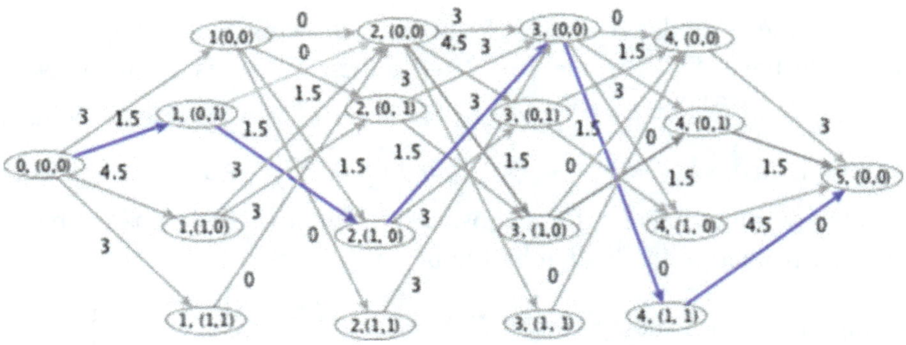

Fig. 2.2 An example of permutation merge

Sect. 2.2.2.2) to match peaks in several profiles simultaneously, based on finding a shortest path in a directed acyclic graph (a graph with directed edges and no cycles). Either Dijkstra algorithm or a topological sort can be used for that Schrijver [20].

Given the n previous profiles, the algorithm builds a directed acyclic graph between each point in time and its predecessors and successors inside a time-window ω, allowing for permutations of values in that window. The cost of each permutation is the difference of the two values that is caused by permutation. Then the minimum weighted path gives an 'averaged' profile with preserved peaks.

As the algorithms complexity is $\mathcal{O}(n\omega^N 4^{N\omega})$, where n is the number of historic profiles, N is the length of time series and ω is time window where permutations are allowed, only small ωs are computationally feasible. If we have two profiles of length five, $\boldsymbol{x} = [0, 0, 3, 0, 0]$ and $\boldsymbol{y} = [3, 0, 0, 0, 0]$ and $\omega = 1$, so we can permute only adjacent values, the constructed graph and the minimum length (weighted) path is given below on Fig. 2.2. As we have two profiles, and are trying to find two permutations that will give us the minimum difference with the median of those two profiles, in each times-step we have 4 possibilities: (0, 0) means both profiles stay the same, (0, 1) the first stays the same, the second is permuted, (1, 0) the first is permuted, the second stays the same, and (1, 1) means both are permuted. As we have to have perfect matching we have $n + 1 = 6$ layers in the graph, and some paths are not available. The solution gives us [0, 3, 0, 0, 0] for both profiles.

2.1.3.3 Adjusted k-nearest Neighbours and Extensions

Valgaev et al. [23] combined the p-adjusted error from Sect. 2.2.2.2 and PM using the k-nearest neighbour (kNN) regression algorithm. The standard kNN algorithm starts by looking for a similar profile in historic data. This is usually done by computing Euclidean based distance between the profiles and returning k minimum distance ones. Then the arithmetic mean is computed and returned as a prediction. Here, instead of the Euclidean distance, the p-adjusted error is used, and instead of computing an arithmetic mean, permutation merge is used to compute adjusted mean. This approach is extended to Adjusted feature aware k-nearest neighbour (AFkNN) in Voß et al. [24] using external factors (temperature, bank holiday, day of week), with one difference. Instead of using adjusted error, the Gower distance

$$D_G(i, j) = \frac{1}{N}\sum_{i=1}^{N} \frac{|x_i^{(f)} - x_j^{(f)}|}{\max x^{(f)} - \min x^{(f)}},$$

is deployed. This is computationally demanding, but can result in better performance than PM in average as it has been shown in Voß et al. [24].

The advantage of permutation based algorithms, as mentioned above, is that these iterative permutations allow forecasted load profiles to look more like the observed load profiles. They are better able to replicate the irregular spikes than the more common averaging or regression based algorithms. However, some error measures such as those that will be discussed in Sect. 2.2.1, can doubly penalise peaky forecasts. Both Charlton et al. [22] and Haben et al. [19] demonstrate how a flat "average" forecast is only penalised once for missing the observed peak whereas if a peak is forecasted slightly shifted from when it actually it occurs, it will be penalised once for missing the peak and again for forecasting it where it was not observed.

2.1.4 Machine Learning Based Algorithms

Machine learning algorithms such as artificial neural networks and support vector machines have been remarkably successful when it comes to understanding power systems, particularly for high voltage systems [25, 26] or aggregated load [2, 9, 27]. The big advantage of machine learning techniques is that they can be quite flexible and are capable of handling complexity and non-linearity [12, 28].

However, the parameters such as weights and biases in a machine learning framework do not always have similarly accessible physical interpretations as in the statistical models discussed, Moreover, some machine learning algorithms such as those used for clustering do not include notions of confidence intervals [29]. Nonetheless, since they have such large scope within and outside of electricity forecasting and since we are mostly interested in point load forecasting in this book, we review two key methods within artificial neural networks, multi-layer perceptron and long short term memory network, and discuss support vector machines.

2.1.4.1 Artificial Neural Networks

Artificial Neural Networks (ANN) are designed to mimic the way the human mind processes information; they are composed of neurons or nodes which send and receive input through connections or edges. From the input node(s) to the output node(s), a neural network may have one or more hidden layers. The learning may be shallow i.e. the network has one or two hidden layers which allows for faster computation.

Or it may be deep, meaning it has many hidden layers. This then allows for more accurate forecasts, but at the cost of time and complexity. When there is a need to forecast many customers individually, computational and time efficiency is a practical requirement for everyday forecasting algorithms. Thus, shallow neural networks such as multi-layer perceptron (MLP) with one hidden layer tended to be used frequently [30–32].

2.1.4.2 Multi-layer Perceptron

MLP is an example of a feedforward ANN. This means that the network is acyclic, i.e. connections between nodes in the network do not form a cycle. MLP consist of three or more layers of nodes: the input layer, at least one hidden layer, and an output layer.

Nodes have an activation function which defines the output(s) of that node given the input(s). In MLP, activation functions tend to be non-linear with common choices being the rectifier ($f(x) = x^+ = \max(0, x)$), the hyperbolic tangent ($f(x) = \tanh(x)$), or the sigmoid function ($f(x) = 1/(1 + e^{-x})$) and the neural network is trained using a method known as backpropagation.

Briefly, it means that the gradient of the error function is calculated first at the output layer and then distributed back through the network taking into accounts the weights and biases associated with the edges and connections in the network. Gajowniczek and Ząbkowski [32] and Zufferey et al. [31] are two recent examples of the use of MLP to individual smart meter data both with one hidden layer.

Gajowniczek and Ząbkowski [32] had 49 perceptrons in the input layer, 38 perceptrons in the hidden layer and the 24 perceptrons in the output layer to coincide with hourly load forecasts. However, Gajowniczek and Ząbkowski [32] was tried on load profile on one household where many details such as occupant number, list of appliances were known. Of course, such information is not freely available.

Zufferey et al. [31], on the other hand, tested a MLP forecast on a larger number of households and found that 200 perceptrons in the hidden layer was a reasonable trade-off between accurate predictions and reasonable computation time. They also found that the inclusion of temperature had limited influence on forecast accuracy, which is similar to the findings of Haben et al. [1] using time-series methods.

2.1.4.3 Long-Short-Term-Memory

Even more recently, Kong et al. [33] used a type of recurrent neural network (RNN) known as the long short-term memory (LSTM) RNN.

These types of models have been successfully used in language translation and image captioning due to the their architecture; since this type of RNN have links pointing back (so they may contain directed cycles, unlike the neural networks discussed before), the decision they make at a past time step can have an impact on the decision made at a later time step.

In this way, they are able to pick up temporal correlation and learn trends that are associated with human behaviour better than traditional feed-forward neural networks. When compared to some naive forecasting algorithms such as the similar day forecast as well as some machine learning algorithms, Kong et al. [33] found that LSTM network was the best predictor for individual load profiles, although with relatively high errors (MAPE, see Sect. 2.2.1, was still about 44% in the best case scenario for individual houses).

The LSTM network that has been implemented in Kong et al. [33] has four inputs: (i) the energy demand from the K past time steps, (ii) time of day for each of the past K energy demand which is one of 48 to reflect half hours in a day, (iii) day of the week which is one of 7, (iv) a binary vector that is K long indicating whether the day is a holiday or not. Each of these are normalised. The energy is normalised using the min-max normalisation.[8]

The normalisation of the last three inputs is done using one hot encoder.[9] The LSTM network is designed with two hidden layers and 20 nodes in each hidden layer. The MAPE (see Sect. 2.2) is lowest for individual houses when LSTM network is used when $K = 12$ though admittedly the improvement is small and bigger improvements came from changing forecasting algorithms.

2.1.4.4 Support Vector Machines

Support Vector Machines (SVM) are another commonly used tool in machine learning, though usually associated with classification. As explained in Dunant and Zufferey [28], SVM classify input data by finding the virtual boundary that separates different clusters using characteristics which can be thought of the features. This virtual boundary is known as the hyper-plane. Of course, there may exist more than one hyper-plane, which may separate the data points. Thus the task of an SVM is to find the hyper-plane such that the distance to the closest point is at a maximum.

We may then use the SVM for regression (SVR) to forecast load as it has been done in Humeau et al. [27] and Vrablecová et al. [34]. In this case, instead of finding the function/hyper-plane that separates the data, the task of the SVR is to find the function that best approximates the actual observations with an error tolerance ϵ. For non-linear solutions, the SVR maps the input data into a higher dimensional feature space.

Humeau et al. [27] used both MLP and SVR to forecast load of single household and aggregate household using data from the Irish smart meter trials. The goal was to create an hour ahead forecast and 24 h ahead forecast. The features used included

[8]Suppose observations are denoted by $\boldsymbol{x}$. These can be normalised with respect to their minimum and maximum values as follows: $z = \frac{\boldsymbol{x}-\min(\boldsymbol{x})}{\max(\boldsymbol{x})-\min(\boldsymbol{x})}$.

[9]The one hot encoder maps each unique element in the original vector that is K long to a new vector that is also K long. The new vector has values 1 where the old vector contained the respective unique element and zeros otherwise. This is done for each unique element in the original vector.

hour of the day and day of the week in calendar variables and the load from the past three hours as well as temperature records in Dublin.

In order to deal with variability and to give an indication of the evolution of load, the authors also add load differences, i.e. $L_i - L_{i-1}$ and $L_i - 2L_{i-1} + L_{i-2}$. Humeau et al. [27] noticed that for individual households, both for the hour ahead and the 24 h ahead, the linear regression outperforms the SVR which the authors do not find surprising. The effectiveness lies in the aggregated load cases where the internal structure, which the SVR is designed to exploit, is clearer.

More recently, Vrablecová et al. [34] also used SVR method to forecast load from the Irish smart meter data. Many kernel functions, which map input data into higher dimensions, were tried. The best results were found using radial basis function kernels and it was noted that sigmoid, logistic and other nonparametric models had very poor results. For individual households, SVR was not found to be the best methods.

Thus, from the reviewed literature we conclude that, while SVR is a promising algorithm of forecasting aggregated load, the volatility in the data reduces its effectiveness when it comes to forecasting individual smart meter data.

2.2 Forecast Errors

As more and more forecasting algorithms become available, assessing how close the forecast is to the truth becomes increasingly important. However, there are many ways to assess the accuracy of a forecast depending on the application, depending on the definition of accuracy and depending on the need of the forecaster. Since one of the earlier comparisons of various forecasting algorithms by Willis and Northcote-Green [35], competitions and forums such as "M-competitions" and "GEFcom" have been used to bring researchers together to come up with new forecasting algorithms and assess their performance. As noted by Hyndman and Koehler [36] and Makridakis and Hibon [37], these competitions help set standards and recommendations for which error measures to use.

2.2.1 Point Error Measures

The **mean absolute percentage error (MAPE)** defined as

$$MAPE = \frac{100\%}{N} \sum_{i=1}^{N} \left| \frac{f_i - a_i}{a_i} \right|, \tag{2.10}$$

where $\boldsymbol{f} = (f_1, \ldots, f_N)$ is the forecast and $\boldsymbol{a} = (a_1, \ldots, a_N)$ is the actual (observed) load, is one of the most common error measures in load forecasting literature.

It is scale independent, so it can be used to compare different data-sets [36]. It is advantageous because it has been historically used and thus often forms a good benchmark. It is also simple and easily interpreted. However, it is not without flaws. As noted in Hyndman and Koehler [36] if $\exists\, i, a_i = 0$, MAPE is undefined. Furthermore, as a point error forecasts, it suffers from the double penalty effect which we shall explain later. In this book, we adopt a common adjustment that allows for data points to be zero and define the MAPE to be

$$MAPE = 100\% \frac{\sum_{i=1}^{N} |f_i - a_i|}{\sum_{i=1}^{N} a_i} \tag{2.11}$$

where $\boldsymbol{f} = (f_1, \ldots, f_N)$ is the forecast and $\boldsymbol{a} = (a_1, \ldots, a_N)$ is the actual (observed) load,

The **Mean Absolute Error (MAE)** is also similarly popular due to its simplicity, although it is scale dependent. We define it as follows

$$MAE = \frac{1}{N} \sum_{i=1}^{N} |f_i - a_i|. \tag{2.12}$$

However, similar to the MAPE, the MAE also susceptible to doubly penalising peaky forecasts. As pointed out in Haben et al. [1], the scale-dependence of MAE can be mitigated by normalising it. In this way the authors were able to compare feeders of different sizes. In our case normalising step is not necessary as we compare different algorithms and look for an average over the fixed number of households.

Haben et al. [19] consider a p-norm error measure. We define it here by

$$E_p \equiv ||\boldsymbol{f} - \boldsymbol{a}||_p := \left(\sum_{i=1}^{N} |f_i - a_i|^p \right)^{\frac{1}{p}}, \tag{2.13}$$

where $p > 1$. In this book, as in Haben et al. [19], we take $p = 4$ as it allows larger errors to be penalised more and smaller errors to be penalised less. Thus, we will use E_4 in order to focus on peaks.

Lastly, we also consider the **Median Absolute Deviation (MAD)** which is defined the median of $|f_i - a_i|$ for $i = 1, \ldots, N$. The MAD is considered more robust with respect to outliers than other error measures.

2.2.2 *Time Shifted Error Measures*

In this section we present some alternatives to the standard error measures listed in Sect. 2.2.1. Suppose the actual profile has just one peak at time k and the forecast also has just one peak at time $k \pm ß$ where $i > 0$ is small (say maybe the next or previous time unit). The point error measures in Sect. 2.2.1 penalise the forecast twice: once for not having the peak at time k and a second time for having a peak at time $k + i$. A flat forecast (fixed value for all time) under these circumstances would have a lower error even though in practice it is not a good or useful to the forecaster. To deal with such problems, it would be intuitively beneficial to be able to associate a shifted peak to a load, including some penalty for a shift, as long as it is within some reasonable time-window. In this section we discuss two ways of comparing time series: dynamic time warping and permuted (so called adjusted) errors.

2.2.2.1 Dynamic Time Warping

Dynamic Time Warping (DTW) is one of the most common ways of comparing two time series not necessarily of equal length. It has been used in automatic speech recognition algorithms.

Dynamic Time Warping calculates an optimal match between two sequences given the some condition: suppose you have time series $\boldsymbol{x}$ and $\boldsymbol{y}$ with length N and M, respectively. Firstly, the index from X can match with more than one index of Y, and vice versa. The first indices must match with each other (although they can match with more than one) and last indices must match with each other.

Secondly, monotonicity condition is imposed, i.e. if there are two indices, $i < j$ for $\boldsymbol{x}$ and if i matches to some index l of $\boldsymbol{y}$, then j cannot match to an index k of $\boldsymbol{y}$ where $k < l$. One can make this explicit: an (N, M)-warping path is a sequence $p = (p1, \dots, p_L)$ with $p_\ell = (n_\ell, m_\ell)$ for $\ell \in \{1, \dots, L\}$, $n_\ell = \{1, \dots, N\}$, and $m_\ell = \{1, \dots, M\}$ satisfying: (i) boundary condition:$p_1 = (1, 1)$ and $p_L = (N, M)$, (ii) monotonicity condition: $n_1 \leq n_2 \leq \dots \leq n_L$ and $m_1 \leq m_2 \leq \dots \leq m_L$ and (iii) step size condition: $p_{\ell+1} - p_\ell \in \{(1, 0), (0, 1), (1, 1)\}$ for $\ell \in \{1, 2, \dots, L - 1\}$.

This is useful to compare electricity load forecast with observation as it allows us to associate a forecast at different time points with the load and still give it credit for having given useful information. However, there are some obvious downsides; the time step in the forecast can be associated with more than one observation which is not ideal. Moreover, if we were to draw matches as lines, lines cannot cross. This means that if we associate the forecast at 7.30 am with observation 8 am, we cannot associate forecast at 8 am with the observation at 7.30 am.

2.2.2.2 Adjusted Error Measure

The adjusted error concept and algorithm to calculate it introduced by Haben et al. [19] addresses some of the issues. The idea of this permutation based error measure was used to create the forecasts discussed in Sect. 2.1.3. Given a forecast and an observation over a time period, both sequences have the same length. Indices are fully (or perfectly) matched. Each index in one sequence is matched only to one index in the other sequence and any permutations within some tolerance window are allowed. The error measure assumes that an observation has been *well enough* forecasted (in time) if both the forecast and the observation are within some time window ω. We define the **Adjusted p-norm Error (ApE)** by

$$\hat{E}_p^{\omega} = \min_{P \in \mathcal{P}} ||P\boldsymbol{f} - \boldsymbol{x}||_p, \tag{2.14}$$

where $\mathcal{P}$ again represents the set of restricted permutations. In Haben et al. [19], the minimisation is done using the Hungarian algorithm, but faster implementations are possible using Dijkstra's shortest path algorithm or topological sort as discussed by Charlton et al. [22].

While these may be intuitively suitable for the application at hand, they are computationally challenging and results are not easily conveyed to a non-specialist audience. The ApE also does not satisfy some properties of metrics In Voß et al. [38], a distance based on the adjusted p-norm error, the local permutation invariant (LPI) is formalised. Let $\mathcal{P}_n$ denote the set of $n \times n$ permutation matrices. Let $\mathcal{L}_n^{\omega} = \{P = (p_{ij} \in \mathcal{P}_n : p_{ij} = 1 \Rightarrow |i - j| \leq \omega\}$. Then the function $\delta : \mathbb{R}^n \times \mathbb{R}^n$, such that

$$\delta(x, y) = \min\{\|Px - y\| : P \in \mathcal{L}_n^{\omega}\}$$

is an LPI distance induced by the Euclidean norm $\|.\|$.

2.3 Discussion

Here, we have looked at several studies that review various load forecasting algorithms and how to assess and compare them. Clearly the literature on how to do this well for individual load profiles is an emerging field. Furthermore, only recently the studies regarding forecasting techniques for individual feeders/households comparing both machine learning algorithms and statistical models became available. This has been done in past for aggregated or high voltage systems [2, 9], but only recently for individual smart meter data.

Linear regression is widely used for prediction, on its own or in combination with other methods. As the energy is used by all throughout the day, and people mostly follow their daily routines, autoregressive models, including ARMA, ARIMA and ARIMAX, SARMA and SARIMAX models are popular and perform well in predicting peaks. Also triple exponential smoothing models, such as Holt-Winters-Taylor with intraday, intraweek and trend components are good contenders, while kernel-density estimators less so for individual households data. As expected, they work better on higher resolutions or aggregated level, where the data is smoother.

Permutation-based methods are relatively recent development. They attempt to mitigate a 'double penalty' issue that standard errors penalise twice slight inaccuracies of predicting peaks earlier or later. Instead of taking a simple average across time-steps, with their adjust averaging they try to obtain a better 'mean sample', and therefore to take into account that although people mostly follow their daily routine, for different reasons their routine may shift slightly in time.

Finally, multi-layer perceptron and recurrent neural network appear to cope well with the volatility of individual profiles, but there is a balance of computational and time complexity and improvement, when comparing them with simpler, explainable and faster forecasts.

There are yet many problems to be solved, such as the question of which features are important factors in individual load forecasting. While in general, there is an agreement that the time of the day, the day of the week and season are important factors for prediction, temperature, which is an important predictor of aggregate level seems to be not very relevant for prediction, (except for households with electric storage heating), due to the natural diversity of profiles being higher than temperature influence.

We have also looked at the standard error measures in order to evaluate different forecasting algorithms. While percentage errors such as are widely used as being scale-free and using absolute values they allow for comparison across different data-sets, we discuss the limitations: a small adjustment allows MAPE to cope with time-series with zero values, but it still suffers from a double penalty problem—trivial, straight line mean forecasts can perform better than more realistic, but imperfect 'peaky' forecasts similarly to MAE. MAD error measure is introduced for error distributions that might be skewed, and 4-norm measure highlights peak errors. Alternatives that use time-shifting or permutations are also mentioned, as they can cope with a double penalty issue, but are currently computationally costly.

References

1. Haben, S., Giasemidis, G., Ziel, F., Arora, S.: Short term load forecasting and the effect of temperature at the low voltage level. Int. J. Forecast. (2018)
2. Alfares, H.K., Nazeeruddin, M.: Electric load forecasting: literature survey and classification of methods. Int. J. Syst. Sci. **33**(1), 23–34 (2002)
3. Taylor, J.W., Espasa, A.: Energy forecasting. Int. J. Forecast. **24**(4), 561–565 (2008)
4. Hong, T., Fan, S.: Probabilistic electric load forecasting: a tutorial review. Int. J. Forecast. **32**(3), 914–938 (2016)
5. Singh, R.P., Gao, P.X., Lizotte, D.J.: On hourly home peak load prediction. In: 2012 IEEE Third International Conference on Smart Grid Communications (SmartGridComm), pp. 163–168 (2012)

6. Taieb, S.B., Taylor, J., Hyndman, R.: Hierachichal probabilistic forecasting of electricity demand with smart meter data. Technical report, Department of Econometrics and Business Statistics, Monash University (2017)
7. Rowe, M., Yunusov, T., Haben, S., Holderbaum, W., Potter, B.: The real-time optimisation of dno owned storage devices on the lv network for peak reduction. Energies **7**(6), 3537–3560 (2014)
8. Gerwig, C.: Short term load forecasting for residential buildings—an extensive literature review. In: Neves-Silva, R., Jain, L.C., Howlett, R.J. (Eds.), Intelligent Decision Technologies, pp. 181–193. Springer International Publishing, Cham (2015)
9. Moghram, I., Rahman, S.: Analysis and evaluation of five short-term load forecasting techniques. IEEE Trans. Power Syst. **4**(4), 1484–1491 (1989)
10. Charlton, N., Singleton, C.: A refined parametric model for short term load forecasting. Int. J. Forecast. **30**(2), 364–368 (2014)
11. Haben, S., Giasemidis, G.: A hybrid model of kernel density estimation and quantile regression for gefcom2014 probabilistic load forecasting. Int. J. Forecast. **32**(3), 1017–1022 (2016)
12. Weron, R.: Modeling and Forecasting Electricity Loads and Prices: A Statistical Approach, vol. 403. Wiley (2007)
13. Taylor, J.W., McSharry, P.E.: Short-term load forecasting methods: an evaluation based on european data. IEEE Trans. Power Syst. **22**(4), 2213–2219 (2007)
14. Box, G., Jenkins, G., Reinsel, G.: Time Series Analysis: Forecasting and Control. Wiley Series in Probability and Statistics. Wiley (2008)
15. Taylor, J.W.: Short-term electricity demand forecasting using double seasonal exponential smoothing. J. Oper. Res. Soc. **54**(8), 799–805 (2003)
16. Arora, S., Taylor, J.W.: Short-term forecasting of anomalous load using rule-based triple seasonal methods. IEEE Trans. Power Syst. **28**(3), 3235–3242 (2013)
17. Taylor, J.W.: Triple seasonal methods for short-term electricity demand forecasting. Eur. J. Oper. Res. **204**(1), 139–152 (2010)
18. Arora, S., Taylor, J.W.: Forecasting electricity smart meter data using conditional kernel density estimation. Omega **59**, 47–59 (2016)
19. Haben, S., Ward, J., Greetham, D.V., Singleton, C., Grindrod, P.: A new error measure for forecasts of household-level, high resolution electrical energy consumption. Int. J. Forecast. **30**(2), 246–256 (2014)
20. Schrijver, A.: Combinatorial Optimization: Polyhedra and Efficiency. Springer, Berlin, Heidelberg (2002)
21. Munkres, J.: Algorithms for the assignment and transportation problems. J. Soc. Ind. Appl. Math. **5**, 32–38 (1957)
22. Charlton, N., Greetham, D.V., Singleton, C.: Graph-based algorithms for comparison and prediction of household-level energy use profiles. In: Intelligent Energy Systems (IWIES), 2013 IEEE International Workshop on, pp. 119–124. IEEE (2013)
23. Valgaev, O., Kupzog, F., Schmeck, H.: Designing k-nearest neighbors model for low voltage load forecasting. In: 2017 IEEE Power Energy Society General Meeting, pp. 1–5 (2017)
24. Voß, M., Haja, A., Albayrak, S.: Adjusted feature-aware k-nearest neighbors: Utilizing local permutation-based error for short-term residential building load forecasting. In: 2018 IEEE International Conference on Communications, Control, and Computing Technologies for Smart Grids (SmartGridComm), pp. 1–6 (2018)
25. Ekonomou, L.: Greek long-term energy consumption prediction using artificial neural networks. Energy **35**(2), 512–517 (2010)
26. Rudin, C., Waltz, D., Anderson, R.N., Boulanger, A., Salleb-Aouissi, A., Chow, M., Dutta, H., Gross, P.N., Huang, B., Ierome, S., et al.: Machine learning for the new york city power grid. IEEE Trans. Pattern Anal. Mach. Intell. **34**(2), 328 (2012)
27. Humeau, S., Wijaya, T.K., Vasirani, M., Aberer, K.: Electricity load forecasting for residential customers: exploiting aggregation and correlation between households. In: Sustainable Internet and ICT for Sustainability (SustainIT), 2013, pp. 1–6. IEEE (2013)

28. Dunant, J., Zufferey, T.: Investigation of forecasting techniques in distribution grids. Semester project of Power System Laboratory, ETHZ (2018)
29. O'Neil, C., Schutt, R.: Doing Data Science: Straight Talk From the Frontline. O'Reilly Media, Inc. (2013)
30. Wijaya, T.K., Vasirani, M., Humeau, S., Aberer, K.: Cluster-based aggregate forecasting for residential electricity demand using smart meter data. In: Big Data (Big Data), 2015 IEEE International Conference on, pp. 879–887. IEEE (2015)
31. Zufferey, T., Ulbig, A., Koch, S., Hug, G.: Forecasting of smart meter time series based on neural networks. In: International Workshop on Data Analytics for Renewable Energy Integration, pp. 10–21. Springer (2016)
32. Gajowniczek, K., Ząbkowski, T.: Short term electricity forecasting using individual smart meter data. Proced. Compu. Sci. **35**, 589–597 (2014)
33. Kong, W., Dong, Z.Y., Jia, Y., Hill, D.J., Xu, Y., Zhang, Y.: Short-term residential load forecasting based on lstm recurrent neural network. IEEE Trans. Smart Grid (2017)
34. Vrablecová, P., Ezzeddine, A.B., Rozinajová, V., Šárik, S., Sangaiah, A.K.: Smart grid load forecasting using online support vector regression. Comput. Electr. Eng. **65**, 102–117 (2018)
35. Willis, H.L., Northcote-Green, J.: Comparison tests of fourteen distribution load forecasting methods. IEEE Trans. Power Appar. Syst. **6**, 1190–1197 (1984)
36. Hyndman, R.J., Koehler, A.B.: Another look at measures of forecast accuracy. Int. J. Forecast. **22**(4), 679–688 (2006)
37. Makridakis, S., Hibon, M.: The m3-competition: results, conclusions and implications. Int. J. Forecast. **16**(4), 451–476 (2000)
38. Voß, M., Jain, B., Albayrak, S.: Subgradient methods for averaging household load profiles under local permutations. In: The 13th IEEE PowerTech 2019 (2019)

Chapter 3
Extreme Value Theory

From travel disruptions to natural disasters, extreme events have long captured the public's imagination and attention. Due to their rarity and often associated calamity, they make waves in the news (Fig. 3.1) and stir discussion in the public realm: is it a freak event? Events of this sort may be shrouded in mystery for the general public, but a particular branch of probability theory, notably Extreme Value Theory (EVT), offers insight to their inherent scarcity and stark magnitude. EVT is a wonderfully rich and versatile theory which has already been adopted by a wide variety of disciplines in a plentiful way. From its humble beginnings in reliability engineering and hydrology, it has now expanded much further; it can be used to model the occurrences of records (say for example in athletic events) or quantify the probability of floods with magnitude greater than what has been observed in the past, i.e it allows us extrapolate beyond the range of available data!

In this book, we are interested in what EVT can tell us about electricity consumption of individual households. We already know a lot about what regions and countries do on average but not enough about what happens at the substation level or at least not with enough accuracy. We want to consider "worst" case scenario such as an area-wide blackout or the "very bad" case scenario such as a circuit fuse blowout or a low-voltage event. Distribution System Operators (DSO) may want to know how much electricity they will need to make available for the busiest time of day up to two weeks in advance. Local councils or policy makers may want to decide if a particular substation is equipped to meet the demands of the residents and if it needs an upgrade or maintenance. EVT can definitely help us to answer some of these questions and perhaps even more as we develop and adapt the theory and approaches further.

There are many ways to infer properties about a population based on various sample statistics. Depending on the statistic, a theory about how well it estimates

M. Jacob et al., *Forecasting and Assessing Risk of Individual Electricity Peaks*, Mathematics of Planet Earth,
https://doi.org/10.1007/978-3-030-28669-9_3

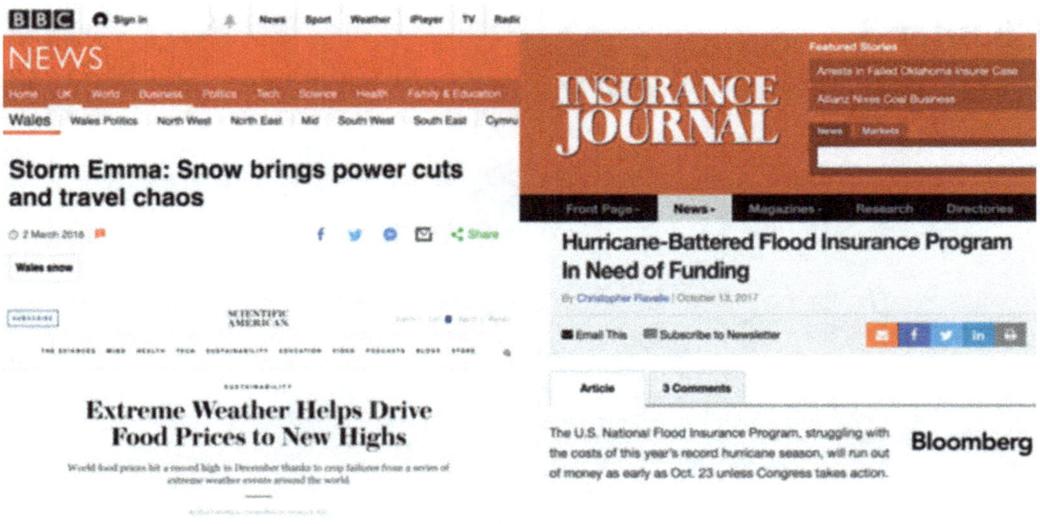

Fig. 3.1 News headlines reporting impacts of various extreme events

the parameter of interest can be developed. The sample average is one such, very common, statistic. Together with the law of large numbers and subsequently the central limit theorem (as well as others), a well known framework exists. However, this framework is lacking in various ways. Some of these limitations are linked to the assumptions of the normal distribution of finite mean and variance. But what if the underlying distribution does not have finite variance or indeed even a finite mean? Not only this, the processes involved in generating a "typical" event may be different to the processes involved in generating an extreme event, e.g. the difference between a windy day and a storm event. Or perhaps extreme events may come from different distributions: for example, internet protocols are the set of rules that set standards on data being transmitted using the internet (or another network). Faloutsos et al. [1] concluded that power-laws can help analyse the average behaviour network protocols whereas simulations from [2] showed exponential distribution in the tails.

EVT establishes the probabilistic and statistical theory of a different sample statistic: unsurprisingly, of extremes. Even though the study of EVT did not gain much traction before [3], some fundamental studies had already emerged in the earlier part of the twentieth century. While not the earliest analysis of extremes, the development of the general theory started with the work of [4]. The paper concerns itself with the distribution of the range of random samples drawn from the normal distribution. It was this work that officially introduced the concept of the distribution of the largest value. In the following years, [5, 6] evaluated the expected value and median of such distributions and the latter extended the question to non-normal distributions. The work of [7–9] gave the asymptotic distributions for the sample extremes. These works summatively give us the extreme value theorem and analogues of the central limit theory for partial or sample maxima. As this is one of the most fundamental results of EVT, we will explore it in more detail in Sect. 3.2.

In essence both the central limit theorem and the extreme value theorem are concerned with describing the same thing; an unusual event. The event may occur as a result from an accumulation of many events or from a single event which exceeds some critical threshold (or not), studied by [10]. Consider, a river whose water levels fluctuate seasonally. These fluctuations may erode its bank overtime or a single flash flooding may break the riverbank entirely. The first is a result of a cumulative effect with which the central limit theorem is concerned whereas the latter is the result of an event which exceeded what the riverbank could withstand, i.e. an extreme event, with which extreme value theory is concerned.

Analogous to measuring "normal" behaviour using a mean, median or a mode, "extreme" behaviour may also be defined in multiple ways. Each definition will give rise to specific results which may be linked to, but different from, results derived from other definitions. However, these results complement each other and allow application to different scenarios/disciplines depending on the nature of the data and the question posed. The subject of EVT is concerned with extrapolation beyond the range of the sample measurements. Hence, it is an asymptotic theory by nature, i.e. results tell us what happens when sample size tends to infinity.

The overarching goal of any asymptotic theory is two fold:

1. to provide the necessary and sufficient conditions to ensure that a specific distribution function (d.f.) occurs in the limit, which are rather qualitative conditions, and
2. to find all the possible distributions that may occur in the limit and derive a generalised form for those distributions.

The first goal is known as the *domains of attraction* problem whereas the second is known as the *limit problem*. Before we take a look at the asymptotic theory, it is valuable to review some of the statistical background and terminology that will be prevalent throughout this report.

The rest of this chapter is dedicated to an in depth exploration of extreme value theory, particularly the probabilistic foundations for the methods of inference presented in Chap. 4 and ensuing application in Chap. 5. In the following section, we will introduce and clarify some of the central terminology and nomenclature. In Sect. 3.2, we will explore the fundamental extreme value theorem which gives rise to the generalised form to the limiting distribution of sample maxima and other relevant results. We will then consider results for the Peaks over threshold (POT) approach in Sect. 3.3. In Sect. 3.4, some theory on regularly varying functions is discussed. The theory of regular variation is quite central to EVT though often it operates from the background. Familiarity with regular variation is highly recommended for the readers interested in a mathematical and theoretical treatment of EVT. Finally, in Sect. 4.4, we consider the case where condition of identically distributed rv's can be relaxed. In Chaps. 3 and 4 as a whole, we aim to provide intuitive understanding of the theoretical results, to expound reasons for these being in great demand to many applications and to illustrate them with some examples of challenges arising in the energy sector.

3.1 Basic Definitions

As was mentioned earlier, various definitions of “extremes” give rise to different, but complementary, limiting distributions. In this section, we want to formalise what we mean by “extreme” as well as introduce the terminology that will be prevalent throughout the book.

Suppose $X_1, X_2, \ldots, X_n, \ldots$ is a sequence of independent random variables (r.v’s). Throughout this book and until said otherwise we shall assume that all these rv’s are generated by the same physical process and therefore it is reasonable to assume that any sample $(X_1, X_2, \ldots, X_n)$, made out of (usually the first) n random variables in the sequence, is a random sample of independent and identically distributed (i.i.d) random variables with common distribution function (d.f.) $F(x) := \mathbb{P}(X \leq x)$, for all $x \in \mathbb{R}$. The non-stationary case, where the underlying d.f. function is allowed to vary with the time or location i of X_i has been worked through in the extreme values context by [11]. We shall refrain to delve into detail on this case within this chapter but in Chap. 4 we shall refer to the statistical methodology for inference on extreme that has sprang from the domain of attraction characterisation furnished by [11]. We shall assume the underlying df is continuous with probability density function (PDF) f. We also often talk about the *support of* X; this is the set of all values of x for which the pdf is strictly positive. The lower endpoint of the support of F or the *left endpoint* is denoted by x_F i.e.,

$$x_F := \inf\{x : F(x) > 0\}.$$

Equivalently, we define the *upper (or right) endpoint* of F by

$$x^F := \sup\{x : F(x) < 1\}. \tag{3.1}$$

which can be either finite or infinite. When each of these values exist finite, these are probabilistically speaking the smallest and largest values that can ever be observed, respectively. In reality we are not likely to observe such extremes or if we do observe extremes we do not know how close they are to the endpoints. This is only to be aggravated by what is generally known in many applications, such as financial or actuarial mathematics, that there might no such thing as a finite upper bound. The main broad aim of EVT is centred around this idea. Its purpose is to enable estimation of tale-telling features of extreme events, right up to the level of that unprecedented extreme event so unlikely to occur that we do not expect it to crop up in the data. Until they do... Now that we have established that EVT sets about teetering on the bring of the sample, aiming to extrapolation beyond the range of the observed data, the sample maximum inherently features as the relevant summary statistic we will be interested in characterising. Preferably with a fully-fledged and complete characterisation, but flexible enough to be taken up by wider applied sciences. A probabilistic result pertaining to the sample maximum such that, for its simplicity and mild assumptions could serve as a gateway for practitioners to find bespoke tail-related models that were not previously accessible or easily interpreted; a theorem like this would be dramatically useful. It turns out that such a result, with resonance often likened to the Central Limit Theorem, already exists. This theorem, the Extreme Value or Extremal Types theorem is the centrepiece to the next section.

Before get underway to the asymptotic (or limit) theory of extremes, we need to familiarise ourselves with the following concepts of convergence for sequences:

- A sequence of random variables $X_1, X_2, \ldots$ is said to *converge in distribution* to a random variable X [notation: $X_n \stackrel{d}{\to} X$] if, for all $x \in \mathbb{R}$,

$$\lim_{n\to\infty} F_n(x) = F(x).$$

  This is also known as weak convergence.
- The sequence *converges in probability* to X [notation: $X_n \stackrel{\mathbb{P}}{\to} X$] if, for any $\epsilon > 0$,

$$\lim_{n\to\infty} P(|X_n - X| > \epsilon) = 0.$$

- The sequence *converges almost surely* to X [notation: $X_n \stackrel{a.s.}{\to} X$] if,

$$P\left(\lim_{n\to\infty} X_n = X\right) = 1.$$

  Almost sure convergence is also referred to as strong convergence.

3.2 Maximum of a Random Sample

In the classical large sample (central limit) theory, we are interested in finding the limit distribution of linearly normalised partial sums, $S_n := X_1 + \cdots + X_n$, where $X_1, X_2, \ldots, X_n \ldots$ are i.i.d. random variables. Whilst the focus here is on the aggregation or accumulation of many observable events, non of these being dominant, the Extreme Value theory shifts to edge of the sample where, for its huge magnitude and potentially catastrophic impact, one single event dominate the aggregation of the data. In the long run, the maximum might not be any less than the sum. Although this seems a bold claim, its probabilistic meaning will become more glaringly apparent later on when we introduce the concept of heavy tails.

The Central Limit theorem entails that the partial sum $S_n = X_1 + X_2 + \ldots + X_n$ of linearly normalised random variables, with constants $a_n > 0$ and b_n, drawn from an iid sequence is asymptotically normal, i.e.

$$\frac{S_n - b_n}{a_n} \xrightarrow[n\to\infty]{d} N(0, 1).$$

reflecting Charlie Winsor's principle that "All actual distributions are Gaussian in the middle". The suitable choice of constant for the Central Limit theorem (CLT) to hold in its classical form are $b_n = nE(X_1)$ and $a_n = \sqrt{nVar(X_1)}$. Therefore, an important requirement is the existence of finite moment of second order, i.e. $E|X_1|^2 < \infty$. This renders the CLT inapplicable to a number of important distributions such as the Cauchy distribution. We refer to [12, Chap. 1] for further aspects on the class of sub-exponential distributions, which includes but is far from limited to the Cauchy distribution.

As it was originally developed, the Extreme Value theorem is concerned with partial maxima $X_{n,n} := \max(X_1, X_2, \ldots, X_n)$ of an iid (or weakly dependent) sequence of rv's. Thus, the related *extremal limit problem* is to find out if there exist constants $a_n > 0$ and b_n such that the limit distribution to $a_n^{-1}(X_{n,n} - b_n)$ is non-degenerate. It is worth highlighting that the sample maximum itself converges *almost surely* to the upper endpoint of the underlying distribution F to the sampled data, for the df of the maximum is given by $P(X_{n,n} \le x) = F^n(x) \to \mathbb{1}_{\{x \ge x^F\}}$, as $n \to \infty$, with $x^F \le \infty$ and $\{X_{n,n}\}_{n \ge 1}$ recognisably a non-decreasing sequence. Here, the indicator function denoted by $\mathbb{1}_A$ is equal to 1 if A holds true and is 0 otherwise, meaning the probability distribution for the maximum distribution has mass confined to the upper endpoint. Hence, the matter now fundamentally lies in answering the following questions: (i) Is is possible to find $a_n > 0$ and b_n such that

$$\lim_{n \to \infty} P\left(\frac{X_{n,n} - b_n}{a_n} \le x\right) = \lim_{n \to \infty} F^n(a_n x + b_n) = G(x), \tag{3.2}$$

for all x continuity points of a non-degenerate cdf G? (ii) What kind of cdf G can be attained in the limit? (iii) How can be G be specified in terms of F? (iv) What are suitable choices for constants a_n and b_n question (i) is referring to? Each of these questions a addresses with great detail in the excellent book by [13].

The celebrated *Extreme Value theorem* gives us the only three possible distributions that G can be. The extreme value theorem (with contributions from [3, 8, 14]) and its counterpart for exceedances above a threshold [15] ascertain that inference about rare events can be drawn on the larger (or lower) observations in the sample. The precise statement is provided in the next theorem. Corresponding result for minima is readily accessible by using the device $X_{1,n} = -\max(-X_1, -X_2, \ldots, -X_n)$.

Theorem 3.1 (Extremal Value Theorem). *Let $\{X_n\}_{n \ge 1}$ be a sequence of i.i.d. random variables with the same continuous df F. If there exist constants $a_n > 0$ and $b_n \in \mathbb{R}$, and some non-degenerate d.f. G such that Eq. (3.2) holds, then G must be only one of three types of d.f.'s:*

Fréchet

$$\Phi_\alpha(x) = \exp\bigl(-x^{-\alpha}\bigr), x > 0 \tag{3.3}$$

Gumbel

$$\Lambda(x) = \exp\bigl(-e^{-x}\bigr), \quad x \in \mathbb{R}, \tag{3.4}$$

Weibull

$$\Psi_\alpha(x) = \exp\big(-(-x)^\alpha\big), x \leq 0 \tag{3.5}$$

The Fréchet, Gumbel and Weibull d.f.'s can be in turn nested in a one-parameter family of distribution through the *von Mises-Jenkinson parameterisation* [16, 17]. Notably, the Generalised Extreme Value (GEV) distribution with df given by

$$G_\gamma(x) = \begin{cases} \exp\big(-(1+\gamma x)^{-1/\gamma}\big), & \gamma \neq 0,\ 1+\gamma x > 0 \\ \exp\big(-e^{-x}\big), & \gamma = 0. \end{cases} \tag{3.6}$$

The parameter $\gamma \in$, the so-called *extreme value index* (EVI), governs the shape of the GEV distribution. In the literature, the EVI is also referred to as the shape parameter of the GEV. We will explore this notion more fully after establishing what it means for F to be in the maximum domain of attraction of a GEV distribution.

Definition 3.1. F said to be in the (maximum-) domain of attraction of G_γ [notation: $F \in \mathcal{D}(G_\gamma)$] if it is possible to redefine constants $a_n > 0$ and b_n provided in Eq. (3.2) in such a way that,

$$\lim_{n\to\infty} F^n(a_n x + b_n) = G_\gamma(x), \tag{3.7}$$

with G_γ given by Eq. (3.6).

We now describe briefly the most salient features in a distribution belonging to each of the maximum domains of attraction:

1. If $\gamma > 0$, then the df F underlying the random sample $(X_1, X_2, \ldots, X_n)$ is said to be in the domain of attraction of a Fréchet distribution with d.f. given by $\Phi_{1/\gamma}$. This domain encloses all df's that are heavy-tailed. Qualitatively, this means that the probability of extreme events are ultimately non-negligible, and that upper endpoint x^F is infinite. Moreover, moments $\mathbb{E}\left(|X|^k\right)$ of order $k > 1/\gamma$ are not finite [18]. Formally, heavy-tailed distributions are those whose tail probability, $1 - F(x)$, is larger than that of an exponential distribution. Thus, noting that $1 - G_\gamma(x) \sim \gamma^{-1/\gamma} x^{-1/\gamma}$, as $x \to \infty$, we can then see that $G_{\gamma>0}$ is a heavy tailed distribution. Pareto, Generalised Pareto, and Inverse Gamma distributions are examples of distributions in the domain of attraction of the Fréchet distribution. Table 2.1 in [19] gives a longer list of distributions in the Fréchet domain of attraction and the corresponding expansion of the tail distribution as well as the EVI in terms of the parameters of the distribution.
2. At a first glance, the Gumbel domain of attraction would be the case of most simple inference as it is obvious that allowing to set $\gamma = 0$, renders any estimation of EVI unecessary. However, the Gumbel domain attracts a plethora of distributions, sliding through a considerable range of tail-heavinesses, from the reversed Fréchet to the Log-Normal, and possessing either finite or infinite upper endpoint. Particular examples and a characterisation of the Gumbel domain can be found in [20].

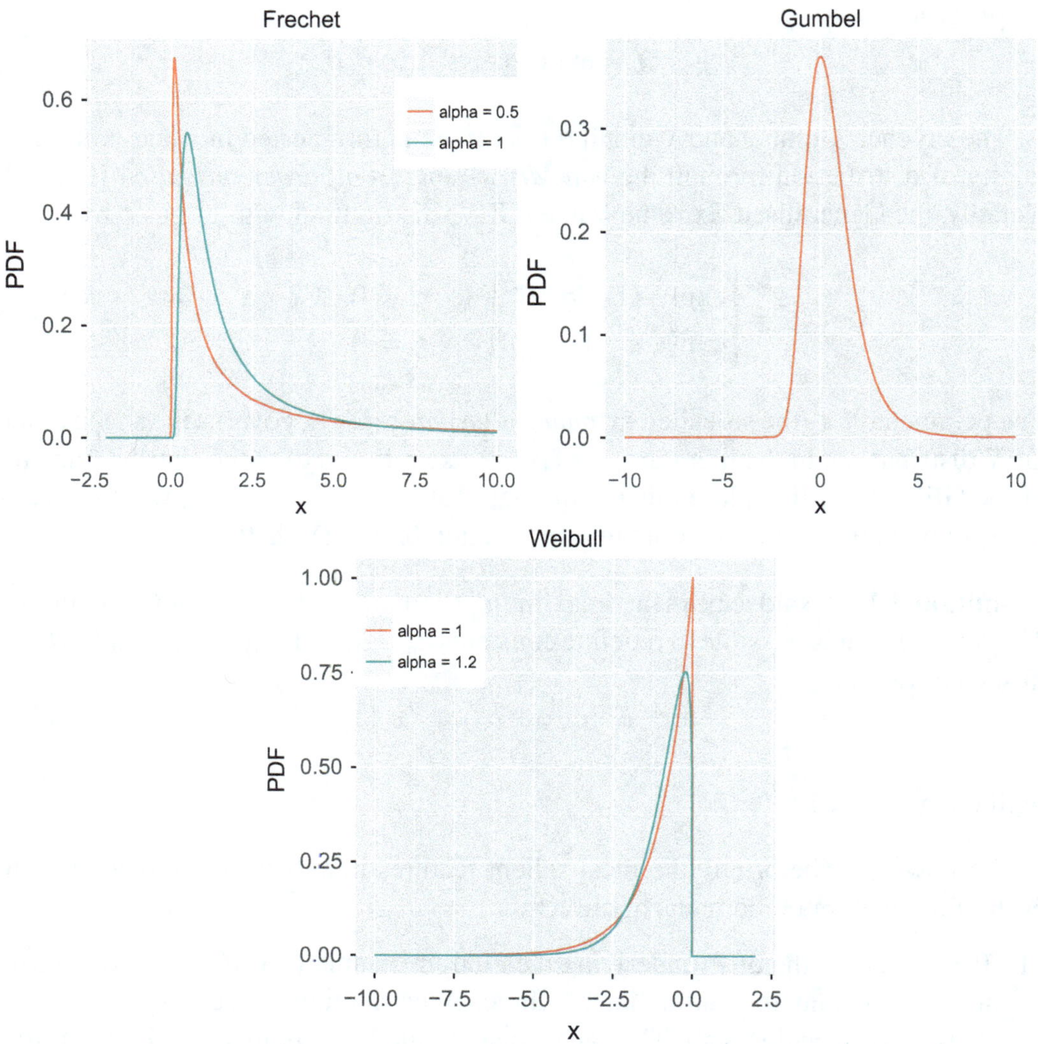

Fig. 3.2 Probability density function of several the extreme value distributions with varying $\alpha = 1/|\gamma|, \alpha > 0$

3. Lastly, for $\gamma < 0$, F is said to be in the domain of attraction of the Weibull distribution with d.f. $\Psi_{-1/\gamma}$. This domain contains short-tailed distributions with finite right endpoint. The case study presented in this report, that of electricity load of individual households, most likely belongs to the Weibull domain of attraction. This is because there is both a contractual obligation for customers to limit their electricity consumption and also physical limit to how much the fuse box can take before it short circuits.

Figure 3.2 displays the prototypical distributions to each domain of attraction, for selected values of α. We highlight the long tails, polynomial decaying tails exhibited by the chosen Fréchet distributions, contrasting with the short, upper bounded, tails ascribed to the Weibull domain.

Now that we have tackled the extremal limit problem, we can start to dissect the domains of attraction. There are enough results on this to publish multiple chapters however not all results are pertinent to our case study. Thus, we take a smaller set of results for the sake of brevity and comprehension. We present the first theorem (Theorem 3.2) which presents a set of equivalent conditions for F to belong to some domain of attraction condition. The proof for this (as for most results) are omitted but can be found in [18] (see Theorem 1.1.6). Theorem 3.2 allows us to make several important observations and connections. As noted before, we have two ways now to see that $F \in \mathcal{D}(G_\gamma)$, one in terms of the tail distribution function $1 - F$ and the other in terms of the tail quantile function U defined as

$$U(t) = \left(\frac{1}{1-F}\right)^{\leftarrow}(t) = F^{\leftarrow}\left(1 - \frac{1}{t}\right), \quad t \geq 1. \tag{3.8}$$

We note that the upper endpoint can thus be viewed as the ultimate quantile in the sense that $x^F = U(\infty) := \lim_{t\to\infty} U(t) \leq \infty$. Secondly, we have possible forms for b_n and some indication as to what a_n might be.

Theorem 3.2. *For $\gamma \in \mathbb{R}$, the following statements are equivalent:*

1. *There exists real constants $a_n > 0$ and $b_n \in \mathbb{R}$ such that*

$$\lim_{n\to\infty} F^n(a_n x + b_n) = G_\gamma(x) = \exp\left(-(1+\gamma x)^{-1/\gamma}\right), \tag{3.9}$$

for all x with $1 + \gamma x > 0$.

2. *There is a positive function a such that for $x > 0$,*

$$\lim_{t\to\infty} \frac{U(tx) - U(t)}{a(t)} = \frac{x^\gamma - 1}{\gamma}, \tag{3.10}$$

where for $\gamma = 0$, the right hand side is interpreted at continuity, i.e. taking the limit as $\gamma \to 0$ giving rise to $\log x$. We use the notation $U \in ERV_\gamma$.

3. *There is a positive function a such that*

$$\lim_{t\to\infty} t[1 - F(a(t)x + U(t))] = (1+\gamma x)^{-1/\gamma}, \tag{3.11}$$

for all x with $1 + \gamma x > 0$.

4. *There exists a positive function f such that*

$$\lim_{t \uparrow x^F} \frac{1 - F(t + xf(t))}{1 - F(t)} = (1+\gamma x)^{-1/\gamma}, \tag{3.12}$$

for all x for which $1 + \gamma x > 0$.

Moreover, a_n and b_n in Eq. (3.9) holds with $a(n)$ and $U(n)$, respectively. Also, $f(t) = a(1/(1 - F(t)))$.

We see in the theorem above where theory of regular variation may come in with regard to U. Though we have deferred the discussion of regular variation to Sect. 3.4, it is useful to note that the tail quantile function U is of extended regular variation (cf. Definition 3.6) and only if F belongs to the domain of attraction of some extreme value distribution. Extreme value conditions of this quantitative nature have resonance from a rich theory and toolbox that we can borrow and apply readily, making asymptotic analysis much more elegant. We can see what the normalising constants might be and we see that we can prove that particular d.f. belongs to the domain of attraction of a generalised extreme value distribution either using F or using U. We may look at another theorem which links the index of regular variations with the extreme value index. Proceeding along these lines, the next theorem borrows terminology and insight from the theory of regular variation; though defined later (Definition 3.4), a slowly varying function ℓ satisfies $\lim_{t\to\infty} \ell(tx)/\ell(x) = 1$. Theorem 3.3 gives us the tail distribution of F, denoted by $\bar{F} = 1 - F$ in terms of a regularly varying function and the EVI. Noticing that $\bar{F}$ is a regularly varying function means we can integrate it using Karamata's theorem (Theorem 3.13) which is useful for formulating functions f satisfying Eq. (3.12).

Theorem 3.3. *Let ℓ be some slowly varying function and $\bar{F}(x) := 1 - F(x)$ denote the survival function, where F is the d.f. association with the random variable X. Let x^F denote the upper endpoint of the df F.*

1. *F is the Fréchet domain of attraction, i.e. $F \in \mathcal{D}(G_\gamma)$ for $\gamma > 0$, if and only if*

$$\bar{F}(x) = x^{-1/\gamma}\ell(x) \iff \lim_{t\to\infty} \frac{1 - F(tx)}{1 - F(t)} = x^{-1/\gamma},$$

for all $x > 0$.

2. *F is in the Weibull domain of attraction, i.e. $F \in \mathcal{D}(G_\gamma)$ for $\gamma < 0$, if and only if*

$$\bar{F}(x^F - x^{-1}) = x^{-1/\gamma}\ell(x) \iff \lim_{t\downarrow 0} \frac{1 - F(x^F - tx)}{1 - F(x^F - t)} = x^{-1/\gamma},$$

for all $x > 0$.

3. *F is in the domain of attraction of the Gumbel distribution, i.e. $\gamma = 0$ with $x^F \leq \infty$*

$$\lim_{t\uparrow x^F} \frac{1 - F(t + xf(t))}{1 - F(t)} = e^{-x},$$

for all $x \in \mathbb{R}$, with f a suitable positive auxiliary function. If the above equation holds for some f, then it also holds with $f(t) := \left(\int_t^{x^F}(1 - F(s))ds\right)/\left(1 - F(t)\right)$ where the numerator of the integral exists finite for $t < x^F$.

Theorem 3.4. *$F \in \mathcal{D}(G_\gamma)$ if and only if for some positive function f,*

$$\lim_{t\uparrow x^F} \frac{1 - F(t + xf(t))}{1 - F(t)} = (1 + \gamma x)^{-1/\gamma} \tag{3.13}$$

for all x with $1 + \gamma x > 0$. *If the above holds for some* f, *then it also holds with*

$$f(t) = \begin{cases} \gamma t, & \gamma > 0 \\ -\gamma(x^F - t), & \gamma < 0 \\ \frac{\int_t^{x^F} 1-F(x)dx}{1-F(t)}, & \gamma = 0. \end{cases}$$

Moreover, any f *that satisfies Eq.* (3.13) *also satisfies*

$$\begin{cases} \lim_{t\to\infty} \frac{f(t)}{t} = \gamma, & \gamma > 0, \\ \lim_{t\uparrow x^F} \frac{f(t)}{x^F - t} = -\gamma, & \gamma < 0, \\ f(t) \sim f_1(t), & \gamma = 0, \end{cases}$$

where $f_1(t)$ *is some function for which* $f_1'(t) \to 0$ *as* $t \uparrow x^F$.

In this section, we have thus far mentioned a suitable function f which plays various roles however it should not be interpreted as probability density function of F, unless explicitly stated as such. Theorem 3.4 gives us alternative forms for f and its limit relations.

Theorem 3.5. *If* $F \in \mathcal{D}(G_\gamma)$, *then*

1. *for* $\gamma > 0$:
$$\lim_{n\to\infty} F^n(a_n x) = \exp\left(-x^{-1/\gamma}\right) = \Phi_{-1/\gamma}(x)$$
holds for $x > 0$ *with* $a_n := U(n)$;
2. *for* $\gamma < 0$:
$$\lim_{n\to\infty} F^n(a_n x + x^F) = \exp\left(-(-x)^{-1/\gamma}\right) = \Psi_{-1/\gamma}(x)$$
for $x < 0$ *holds with* $a_n := x^F - U(n)$;
3. *for* $\gamma = 0$:
$$\lim_{n\to\infty} F^n(a_n x + b_n) = \exp\left(-e^{-x}\right) = \Lambda(x)$$
holds for all x *with* $a_n := f(U(n))$ *and* $b_n := U(n)$ *where* f *is as defined in Theorem 3.3 (3).*

We briefly consider maxima that have not been normalised and have been normalised for various sample sizes $n = 1, 7, 30, 365, 3650$ (Fig. 3.3). The left plot of Fig. 3.3 shows the distribution of the maxima where the lines represent, from left to right, each of the sample sizes. The right hand side of the same plot shows how quickly, the normalised maxima go to the Gumbel distribution; the exponential distribution belongs to the Gumbel domain of attraction. The appropriate normalising constants for F standard exponential are $a_n = 1$ and $b_n = \log(n)$. Deriving this is left as an exercise.

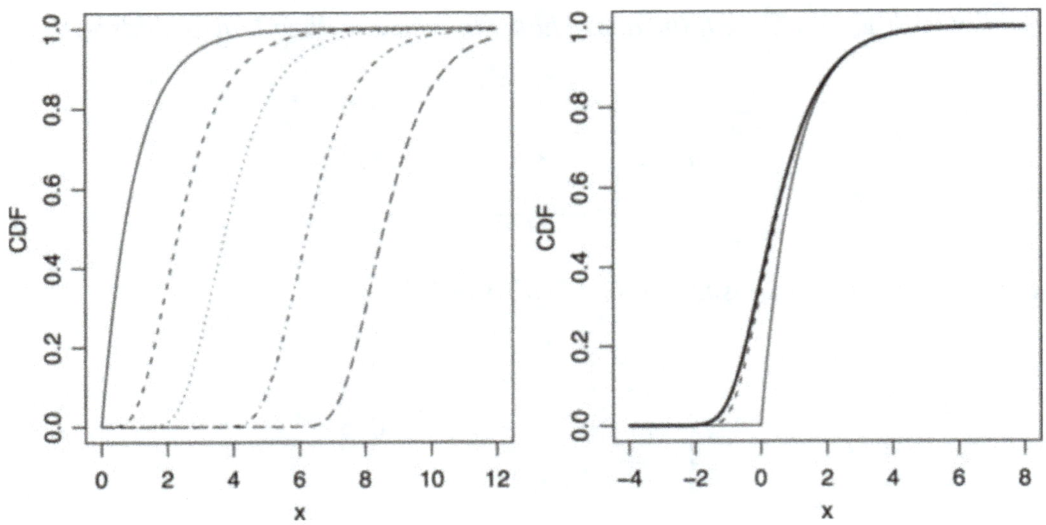

Fig. 3.3 Distributions of maxima (left) and normalised maxima (right) of $n = 1, 7, 30, 365, 3650$ standard exponential random variables, with the Gumbel d.f. (solid heavy line)

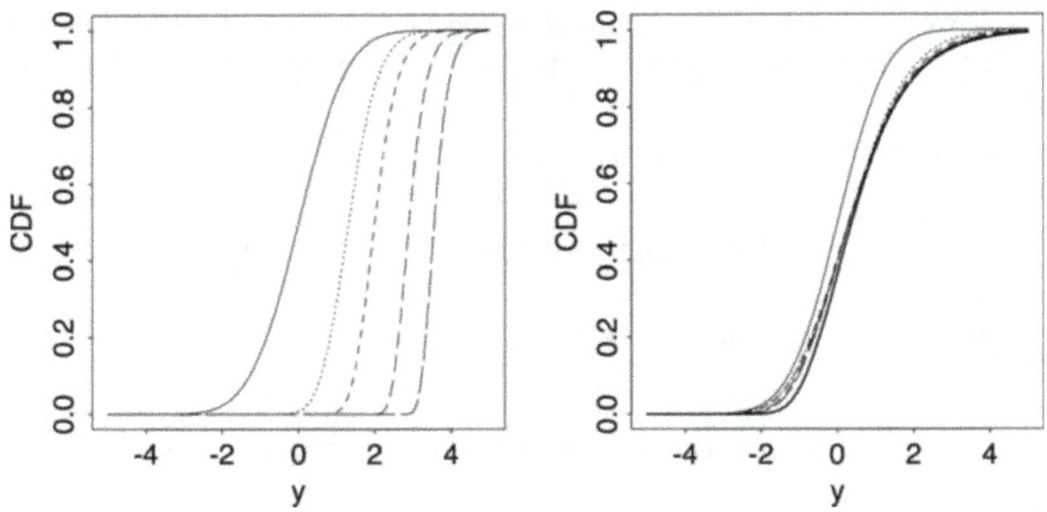

Fig. 3.4 Distributions of maxima (left) and normalised maxima (right) of $n = 1, 7, 30, 365, 3650$ standard normal random variables, with the Gumbel d.f. (solid heavy line)

Doing the same thing except now with F standard normal distribution, Fig. 3.4 shows that the convergence is slow. In this case, $a_n = (\log n)^{-1/2}$ and $b_n = (\log n)^{1/2} - 0.5\,(\log n)^{-1/2}\,(\log\log n + \log 4\pi)$. As before, F standard normal belongs to the Gumbel domain of attraction.

Theorem 3.5 gives which sequences a_n and b_n should be used to normalise maxima in order to ensure that F is in the maximum domain of attraction of a specific G_γ. Note that the values of a_n and b_n changes with the sample size, n. If γ were known before hand, knowing the true value of the normalising constants may help with simulations or numerical experiments. However in practice, we do not know γ and it must be estimated. Thus, we can use the von Mises condition give us a work around.

Theorem 3.6 (von Mises Condition). *Let $r(x)$ be the hazard function defined by*

$$r(x) = \frac{f(x)}{1 - F(x)} \tag{3.14}$$

where $f(x)$ is the probability density function and F is the corresponding d.f..

1. *If $x^F = \infty$ and $\lim_{x \uparrow \infty} x\, r(x) = 1/\gamma > 0$, then $F \in \mathcal{D}(\Phi_{1/\gamma})$.*
2. *If $x^F < \infty$ and $\lim_{x \uparrow x^F} (x^F - x)\, r(x) = -1/\gamma > 0$, then $F \in \mathcal{D}(\Psi_{-1/\gamma})$.*
3. *If $r(x)$ is ultimately positive in the negative neighbourhood of x^F, is differentiable there and satisfies $\lim_{x \uparrow x^F} \frac{d}{dx} r(x) = 0$, then $F \in \mathcal{D}(\Lambda)$.*

The von Mises conditions given in Theorem 3.6 is particularly useful when one is interested in conducting simulations. We may sample from a known distribution F which readily gives us the probability density function, f. Thus, without knowledge of the appropriate normalising constants, the von Mises conditions allow us to find the domain of attraction of F.

We have discussed the asymptotic theory of the maximum from a sample. Earlier we mentioned that in practice, we divide the data into blocks of length n and take the maximum from each block and conduct inference on them. The results we have discussed in this section, tell us what happens as the block size becomes infinitely large. The approach of sampling maxima from blocks is, unsurpsingly known as the *Block Maxima* approach. As [21] pointed out, the block maxima model offers many practical advantages (over the Peaks Over Threshold, Sect. 3.3). The block maxima method is the appropriate statistical model when only the most extreme data are available; for example, historically temperature data was recorded in daily minimum, average and maximum. In cases where the time series may have periodicity, we can remove some dependence by dividing the blocks in such a way that dependence may exist within the block but not between blocks. We will now consider an alternative but equivalent method.

3.3 Exceedances and Order Statistics

When conducting inference on the tail of a distribution, it is wasteful to consider only the most extreme observation. We may be able to glean valuable information by utilising more than just the maximum. For such cases, we may study either exceedances over a (high) threshold (Sect. 3.3.1) or we may consider order statistics (Sect. 3.3.2). In each case, we get different limiting distributions. In what follows we will discuss what the limiting distributions are in each case and how they relate to the extreme value distributions and the results from Sect. 3.2.

3.3.1 Exceedances

In this instance, the idea is that statistical inference is be based on observations that exceed a high threshold, u, i.e., either on X_i or on $X_i - u$ provided that $X_i > u$ for $i \leq n$. The exact conditions under which the POT model hold is justified by second order refinements (cf. Sect. 3.4) whereas typically it has been taken for granted that the block maxima method follows the extreme value distribution very well. We saw this from the discussion from Fig. 3.4. For large sample sizes, the performance of the block maxima method and the peaks over threshold method is comparable. However, when the sample is not large, there may be some estimations where the Peaks over threshold (POT) model is more efficient [21].

Since we have familiarised ourself with the convergence of partial maxima, we now do the same for exceedance. We will show that appropriately normalised exceedances converge to the Generalised Pareto distribution. This is the POT model. The work on exceedances was independently initiated by [15, 22]. As before, we will start with definitions and then proceed to establishing the Generalised Pareto as the limiting distribution.

Definition 3.2. Let X be a random variable with d.f. F and right endpoint x^F. Suppose we have the threshold $u < x^F$. Then the d.f., F_u, of the random variable X over the threshold u is defined to be

$$F_u(x) = \mathbb{P}(X - u \leq x | X > u) = \frac{F(x+u) - F(u)}{1 - F(u)}, \quad 0 \leq x < x^F - u. \tag{3.15}$$

Definition 3.3. The *Generalised Pareto* distribution is defined as

$$W_\gamma(x) = \begin{cases} 1 - (1 + \gamma x)^{-1/\gamma}, & \gamma \neq 0, \\ 1 - e^{-x}, & \gamma = 0 \end{cases} \tag{3.16}$$

where

$$\begin{cases} x \geq 0, & \gamma \geq 0, \\ 0 \leq x \leq -1/\gamma, & \gamma < 0. \end{cases}$$

Note that, as for G_γ, the Generalised Pareto distribution also has scale and location parameters: $W_{\gamma;\,\nu,\,\beta}(x) := W_\gamma((x - \nu)/\beta)$. Further, for x with $1 + \gamma x > 0$,

$$1 + \log G_\gamma(x) = W_\gamma(x).$$

Now that we have looked at the d.f. of exceedance and defined the Generalised Pareto distribution, the latter can be established as the limiting distribution of exceedances.

Theorem 3.7 (Balkema-de Haan-Pickands Theorem). *One can find a positive, measurable function* β *such that*

$$\lim_{u \uparrow x^F} \sup_{0 \le x \le x^F - u} |F_u(x) - GP_{\gamma;\,0,\beta(u)}(x)| = 0 \tag{3.17}$$

if and only if $F \in \mathcal{D}(G_\gamma)$.

Not only does the Balkema-de Haan-Pickands theorem allow us to use more than the maximum, it also connects the d.f. of the random variables to that of exceedances over a threshold; from knowing the limiting distribution of F_u, we also know about the domain of attraction of F and vice versa. The shape parameter in both cases is the same and thus their interpretation is the same as before, i.e., γ describes the tail-heaviness of F if Eq. (3.17) is satisfied. Holmes and Moriarty [23] used the Generalised Pareto distribution to model particular storms of interest for applications in wind engineering and [24] used the POT method to analyse financial risk.

3.3.2 Asymptotic Distribution of Certain Order Statistics

In the previous section, we talked about how the POT approach can use data more efficiently. The efficiency relies on choosing the threshold appropriately. If the threshold is too low, then the exceedances are no longer from the tail and the bias is dominant. On the other hand, if the threshold is too high, then very few data points exceed it and the variance is high and confidence in the results is low. We can consider this idea of balancing the effects of bias and variance by considering a certain kind of order statistics. This is the topic of this section.

Suppose $X_1, X_2, \ldots, X_n, \ldots$ are i.i.d. random variables with common d.f. F. If we take a finite sample $X_1, \ldots, X_n$ and order it from minimum to maximum, then we get the nth *order statistics*:

$$X_{1,n} \le X_{2,n} \le \cdots \le X_{n,n}. \tag{3.18}$$

Furthermore, we can define the kth *upper order statistic*, $X_{n-k,n}$, to be the kth largest value in the finite sample; the nth upper order statistic, i.e. $k = n$ is the maximum and the first upper order statistic, i.e. $k = 1$, is the minimum. Depending on k and its relation to n, the kth upper order statistic can be classified in at least three different ways which leads to different asymptotic distributions. Arnold et al. [13] classified $X_{n-k,n}$ to be one the following three order statistics:

1. *Central Order Statistics*: $X_{n-k,n}$ is considered to be a central order statistic if $k = [np] + 1$ where $0 < p < 1$ and $[\cdot]$ characterises the function which is the smallest integer larger than the argument.
2. *Extreme Order Statistics*: $X_{n-k,n}$ is an extreme order statistic when either k or $n - k$ is fixed and $n \to \infty$.

3. *Intermediate Order Statistics*: $X_{n-k,n}$ is an intermediate sequence if both k and $n-k$ approach infinity but $k/n \to 0$ or 1. In this book we present results for $k/n \to 0$ and we also assume that k varies with n i.e. $k = k(n)$.

Note that the conditions which ensure that $X_{n-k,n}$ is an intermediate order statistic has similar notions of balancing bias and variance; insisting that $k/n \to 0$ means that all data points larger than $X_{n-k,n}$ is a small part of the entire population and ensures analyses pertains to the tail of the distribution. However, for asymptotic results to hold, some of which we have seen in Sects. 3.2 and 3.3.1 and will see in this section, we require a large enough sample, i.e. k should go to infinity. As such identifying k appropriately is a crucial and a non-trivial part of extreme value analysis and also proves useful for the POT model as it allows us to chose u to be value which corresponds to the intermediate order statistics, $X_{n-k,n}$.

Since we use intermediate order statistics in our case study on electricity load in Chap. 5, it is of more immediate interest to us but for the sake of completeness and intuitive understanding we discuss the asymptotic distribution of all three order statistics. First, we consider the convergence of the kth upper order statistics.

Theorem 3.8. *Let F be a d.f. with a right (left) endpoint $x^F \le \infty$ $(x_F \ge -\infty)$ and $k = k(n)$ be a non-decreasing integer sequence such that*

$$\lim_{n\to\infty} \frac{k(n)}{n} = c \in [0, 1].$$

1. *Then $X_{n-k(n),n} \overset{a.s.}{\to} x^F$ (x_F) as $c = 0$ $(c = 1)$.*
2. *If we instead assume that $c \in (0, 1)$ is such that there is unique solution $x(c)$ of the equation $\bar{F}(x) = c$. Then*

$$X_{n-k(n),n} \overset{a.s.}{\to} x(c).$$

Note that result 3.8 of Theorem 3.8 relates to intermediate order statistics whereas result 3.8 relates to central order statistics. The proof is simple and can be found in [12]. We now proceed to the discussion of the asymptotic distribution for each of the order statistics.

Theorem 3.9 (Asymptotic distribution of a central order statistic). *For $0 < p < 1$, let F be an absolutely continuous d.f. with density function f which is positive at $F^{\leftarrow}(p)$ and is continuous at that point. For $k = [np] + 1$, as $n \to \infty$,*

$$\sqrt{n} f(F^{\leftarrow}(p)) \frac{X_{n-k,n} - F^{\leftarrow}(p)}{\sqrt{p(1-p)}} \overset{d}{\to} \mathcal{N}(0, 1), \tag{3.19}$$

where $\mathcal{N}(\mu, \sigma^2)$ denotes a normal distribution with mean μ and variance σ^2. Thus note that the central order statistics, when appropriately normalised, converges to the normal distribution. This property is known as *asymptotic normality* and is particularly desirable for estimators as it allows for the construction of confidence intervals with relative ease. The proof of Theorem 3.9 can be found in [13].

We can consider the asymptotic distribution of the extreme order statistics (also known as the upper order statistic) which no longer exhibits asymptotic normality. Instead, in this case, we recover links to the Generalized Extreme Value distribution, G.

Theorem 3.10 (Asymptotic distribution of an extreme order statistic). *For any real* x, $\mathbb{P}(X_{n,n} \leq a_n x + b_n) \to G(x)$ *as* $n \to \infty$ *if and only for any fixed* k,

$$\mathcal{P}(X_{n-k+1,n} \leq a_n x + b_n) \overset{n\to\infty}{\longrightarrow} \sum_{j=0}^{k-1} G(x) \frac{(-\log G(x))^j}{j!}, \tag{3.20}$$

for all x.

The proof can be found in [13]. Note that F being in the domain of attraction of an extreme value distribution implies that Eq. (3.20) holds with the same a_n and b_n and thus establishes a strong link between the asymptotic behaviour of extreme order statistics and the sample maxima. However, when k is allowed to vary with n as for intermediate order statistics, we again acquire asymptotic normality.

Theorem 3.11 (Asymptotic distribution of an intermediate order statistic). *Suppose the von Mises condition given in Theorem 3.6 holds for some* G_γ. *Suppose further that* $k \to \infty$ *and* $k/n \to 0$ *as* $n \to \infty$. *Then*

$$\sqrt{k} \frac{X_{n-k,n} - U(\frac{n}{k})}{\frac{n}{k} U'(\frac{n}{k})} \overset{d}{\to} \mathcal{N}(0, 1). \tag{3.21}$$

A proof for Theorem 3.11 can be found in [25]. Thus we see that although intermediate order statistics is somewhere between central order statistics and extreme order statistics and intuitively closer to the latter, its asymptotic behaviour is more akin to that of the central order statistics. Theorem 3.11 also gives us the appropriate normalisation. We now consider an example as it will demonstrate how to use Theorems 3.9 and 3.11. It is also useful for numerical simulation.

Similarly, Theorems 3.9 and 3.10 can be used to choose appropriate normalisation for the relevant order statistics. Of course, in the above example, we have readily applied Theorem 3.11. In practice, we will need to check the von Mises condition or other relevant assumptions. This is taken for granted in the above example.

Order statistics are particularly useful as they are used to build various estimators for γ and x^F. The commonly used Hill estimator for $\gamma > 0$, is an example as is the more general Pickands estimator.

3.4 Extended Regular Variation

We have already alluded to the topics in this section however due to the technical complexity, it is given only at the end. The theory of regular variation provides us a tool box for understanding various functions that we have come across. Moreover, to set the theory that we have discussed within a wider framework, stronger conditions are necessary. These conditions follow readily if we are familiar with the theory of regular variation. The topics in this section may seem disjointed and irrelevant but in fact, it is instrumental to making extreme value theory as rich and robust as it is. We will start with the fundamentals.

Definition 3.4. Let ℓ be an eventually positive function on $\mathbb{R}_+$. Then ℓ is said to be *slowly varying* if and only if

$$\lim_{u\to\infty}\frac{\ell(ux)}{\ell(x)} = 1.$$

Similarly, we can offer a more general version as follows.

Definition 3.5. Let f be an eventually positive function on $\mathbb{R}_+$. Then f is said to be *regularly varying* with index ρ if and only if there exists a real constant ρ such that

$$\lim_{u\to\infty}\frac{f(ux)}{f(x)} = x^{\rho}. \tag{3.22}$$

ρ is called the *index of regular variation* [notation: $f \in RV_\rho$]. Note that if f satisfies Eq. (3.22) with $\rho = 0$, then f is slowly varying. Strictly speaking the above definitions require $f : \mathbb{R}_+ \to \mathbb{R}$ to be Lebesgue measurable. We can readily assume this as most functions in our case are continuous and thus satisfy Lebesgue measurability. Note also that all regularly varying functions f can be written in terms of the a slowly varying function ℓ, i.e., if $f \in RV_\rho$, then $f(x) = x^{\rho}\ell(x)$ where $\ell \in RV_0$. Note then that in Theorem 3.3, the tail of F was regularly varying in both the Fréchet and Weibull cases.

We can make this even more general by considering functions that are of *extended regular variation* and/or belonging to a class of functions denoted by Π.

Definition 3.6. A measurable function $f : \mathbb{R}_+ \to \mathbb{R}$ is said to be of extended regular variation if there exists a function $a : \mathbb{R}_+ \to \mathbb{R}_+$ such that for some $\alpha \in \mathbb{R}\backslash\{0\}$ and for $x > 0$,

$$\lim_{u\to\infty}\frac{f(ux) - f(x)}{a(x)} = \frac{x^{\alpha} - 1}{\alpha}. \tag{3.23}$$

[Notation: $f \in ERV_\alpha$]. The function a is the *auxiliary function* for f. While we do not show this, $a \in RV_\alpha$. We can see now observe that $F \in \mathcal{D}(G_\gamma) \implies U \in ERV_\gamma$ with auxiliary function $a(t)$ (cf. Theorem 3.2). Not only this but we can link f to be regularly varying as follows.

Theorem 3.12. *Suppose f holds with Eq. (3.23), with $\alpha \neq 0$. Then*

1. *If $\alpha > 0$, then $\lim_{x\to\infty} f(x)/a(x) = 1/\alpha$ and hence $f \in RV_\alpha$.*
2. *If $\alpha < 0$, then $f(\infty) := \lim_{x\to\infty} f(x)$ exists, $\lim_{x\to\infty}(f(\infty) - f(x))/a(x) = -1/\alpha$ and hence $f(\infty) - f(x) \in RV_\alpha$.*

The proof can be found in Appendix B of [18]. Since we now have the relation between the normalising constants and EVI with the index of regular variation, it can be used to construct estimators for the EVI. It can also be used in simulations where the true value is known or can be calculated.

Definition 3.7. A measurable function $f : \mathbb{R}_+ \to \mathbb{R}$ is said to belong to the class Π if there exist a function $a : \mathbb{R}_+ \to \mathbb{R}_+$ such that, for $x > 0$,

$$\lim_{u\to\infty} \frac{f(ux) - f(x)}{a(x)} = \log x,$$

where a is again the auxiliary function for f [Notation: $f \in \Pi$ or $f \in \Pi(a)$]. In this case, a is measurable and slowly varying. Note that functions that belong to class Π are a special case of functions which are of extended regular variation, i.e. where the index is 0. Next we consider *Karamata's theorem* which gives us a way to integrate regularly varying function.

Theorem 3.13 (Karamata's Theorem). *Suppose $f \in RV_\alpha$. There exists $t_0 > 0$ such that $f(t)$ is positive and locally bounded for $t \geq t_0$. If $\alpha \geq -1$, then*

$$\lim_{t\to\infty} \frac{tf(t)}{\int_{t_0}^{t} f(s)ds} = \alpha + 1. \tag{3.24}$$

If $\alpha < -1$, or $\alpha = -1$ and $\int_0^\infty f(s)ds < \infty$, then

$$\lim_{t\to\infty} \frac{tf(t)}{\int_t^\infty f(s)ds} = -\alpha - 1. \tag{3.25}$$

Conversely, if Eq. (3.24) holds with $-1 < \alpha < \infty$, then $f \in RV_\alpha$. If Eq. (3.25) holds with $-\infty < \alpha < -1$, then $f \in RV_\alpha$.

Note that in Theorem 3.13, the converse for $\alpha = -1$ does not necessarily imply that f is regularly varying.

It is obvious how the definitions and theorems we have looked at so far are relevant; we have provided examples of functions that were used in the report that satisfy one or more definition. Recall that in Sect. 3.3, we made mentions of second order refinements. The next part, though glance rather terse at first glance, provides a good source of valuable information to the prediction of distinctive features in extreme data. We shall look further at extended regular variation of U in Eq. (3.10) (i.e., Eq. (3.6) specialised in U) to give thorough insight as to how the normalised spacings of quantiles attain the GPD tail quantile function in the limit. The second order refinement below seeks to address the order of convergence in Eq. (3.10).

Definition 3.8. The function U is said to satisfy the *second order refinement* if for some positive function a and some positive or negative function A with $\lim_{t\to\infty} A(t) = 0$,

$$\lim_{t\to\infty} \frac{\frac{U(tx)-U(t)}{a(t)} - \frac{x^\gamma - 1}{\gamma}}{A(t)} =: H(x), \qquad x > 0, \tag{3.26}$$

where H is some function that is not a multiple of $D_\gamma := (x^\gamma - 1)/\gamma$.

The non-multiplicity condition is merely to avoid trivial results. The functions a and A may be called the first-order and second-order auxiliary functions, respectively. As before, the function A controls the speed of convergence in Eq. (3.10). The next theorem establishes the form of H and gives some properties of the auxiliary functions.

Theorem 3.14. *Suppose the second order refinement Eq. (3.26) holds. Then there exist constants c_1, $c_2 \in \mathbb{R}$ and some parameter $\rho \le 0$ such that*

$$H_{\gamma,\rho}(x) = c_1 \int_1^x s^{\gamma-1} \int_1^s u^{\rho-1} du\, ds + c_2 \int_1^x s^{\gamma+\rho-1} ds. \tag{3.27}$$

Moreover, for $x > 0$,

$$\begin{aligned} \lim_{t\to\infty} \frac{\frac{a(tx)}{a(t)} - x^\gamma}{A(t)} &= c_1 x^\gamma \frac{x^\rho - 1}{\rho}, \\ \lim_{t\to\infty} \frac{A(tx)}{A(t)} &= x^\rho. \end{aligned} \tag{3.28}$$

The results are understood in continuity i.e. taking the limit as ρ and/or γ goes to zero. This gives us that

$$H_{\gamma,\rho}(x) = \begin{cases} \frac{c_1}{\rho}\left(D_{\gamma+\rho}(x) - D_\gamma(x)\right) + c_2 D_{\gamma+\rho}(x), & \rho \ne 0, \gamma \ne 0 \\ \frac{c_1}{\gamma}\left(x^\gamma \log x - D_\gamma(x)\right) + c_2 D_\gamma(x), & \rho = 1, \gamma \ne 0 \\ \frac{c_1}{2}(\log x)^2 + c_2 \log x, & \rho = 0, \gamma = 0. \end{cases} \tag{3.29}$$

The case of $\gamma = 0$ and/or $\rho = 0$ is interpreted in the limiting sense as $\log x$. Without loss of generality the constants featuring in the above can set fixed at $c_1 = 1$ and $c_2 = 0$ (cf. Corollary 2.3.4 of [18]). The parameter ρ describes the speed of convergence in Eq. (3.26): ρ close to zero implies slow convergence whereas if $|\rho|$ large, then convergence is fast. The above theorem results from the work of [26]. Finally, we can provide the sufficient second order condition of von Mises type.

Theorem 3.15. *Suppose the tail quantile function U is twice differentiable. Define $A(t) := \frac{tU''(t)}{u'(t)} - \gamma + 1$. If A has constant sign for large t, $\lim_{t\to\infty} A(t) = 0$, and if $|A| \in RV_\rho$ for $\rho \le 0$, then for $x > 0$,*

$$\lim_{t\to\infty} \frac{\frac{U(tx)-U(t)}{tU'(t)} - \frac{x^{\gamma}-1}{\gamma}}{A(t)} = H_{\gamma,\rho}(x).$$

These definitions and results may seem unrelated or arbitrary but in fact some of the proofs of other results borrow understanding from the theory of regular variation, and functions such as the tail quantile function U as seen as of extended regular variation. Thus, regular variation theory allows us to extend the theory of extremes much further in a very natural way, it enables a full characterisation at high levels of the process generating the data by looking at the asymptotic behaviour of the exceedances above a sufficiently high threshold. It also allows us to prove asymptotic normality for various estimators. Thus, though quite involved, it is a very useful tool in extreme value analyses and is highly recommended for the enthusiastic or mathematically motivated reader.

In conclusion, extreme value theory gives us a broad and well grounded foundation to extrapolate beyond the range of available data. Using either sample maxima or exceedances over a threshold, valuable inferences about extremes can be made. These are made rigorous by the first order and second order conditioning, which are underpinned by the broader still theory of regular variation. Moreover, we have techniques to conduct these analyses even when conditions of independence and stationarity do not hold. These results have already been adapted to fields such as finance, flood forecasting, climate change. They are accessible to yet more fields, and in this book they will be adapted for electricity demand in low-voltage networks.

References

1. Faloutsos, M., Faloutsos, P., Faloutsos, C.: On power-law relationships of the internet topology. In: ACM SIGCOMM Computer Communication Review, vol. 29, pp. 251–262. ACM (1999)
2. Yook, S.-H., Jeong, H., Barabási, A.-L.: Modeling the internet's large-scale topology. Proc. Natl. Acad. Sci. **99**(21), 13382–13386 (2002)
3. Haande Haan, L.: On regular variation and its application to the weak convergence of sample extremes. Ph.D. thesis, Mathematisch Centrum Amsterdam (1970)
4. Bortkiewiczvon Bortkiewicz, L.: Variationsbreite und Mittlerer Fehler. Berliner Mathematische Gesellschaft (1922)
5. Misesvon Mises, L.: Über die Variationsbreite einer einer Beobachtungsreihe. Sitzungsberichte der Berliner Mathematischen Gesellschaft **22**, 3–8 (1923)
6. Dodd, E.L.: The greatest and the least variate under general laws of error. Trans. Am. Math. Soc. **25**(4), 525–539 (1923)
7. Fréchet, M.: Sur la loi de probabilité de l'écart maximum. Annales de la Société Polonaise de Mathematique 93–117 (1927)
8. Fisher, R.A., Tippett, L.H.C.: Limiting forms of the frequency distribution of the largest or smallest member of a sample. In: Mathematical Proceedings of the Cambridge Philosophical Society, vol. 24, pp. 180–190. Cambridge University Press (1928)
9. Gnedenko, B.V.: On a local limit theorem of the theory of probability. Uspekhi Matematicheskikh Nauk **3**(3), 187–194 (1948)
10. Gumbel, E.: Statistics of Extremes, p. 247. Columbia University Press, New York (1958)
11. Haande Haan, L.: Convergence of heteroscedastic extremes. Stat. Probab. Lett. **101**, 38–39 (2015)
12. Embrechts, P., Klüppelberg, C., Mikosch, T.: Modelling Extremal Events for Insurance and Finance. Springer, Berlin (1997)

13. Arnold, B.C., Balakrishnan, N., Nagaraja, H.N.: A First Course in Order Statistics. Wiley Series in Probability and Mathematical Statistics. Wiley (1992)
14. Gnedenko, B.: Sur la distribution limite du terme maximum d'une serie aleatoire. Ann. Math. 423–453 (1943)
15. Balkema, A.A., Haande Haan, L.: Residual life time at great age. Ann. Probab. 792–804 (1974)
16. Misesvon Mises, R.: La distribution de la plus grande de n valeurs. Rev. math. Union interbalcanique **1**(1) (1936)
17. Jenkinson, A.F.: The frequency distribution of the annual maximum (or minimum) values of meteorological elements. Q. J. R. Meteorol. Soc. **81**(348), 158–171 (1955)
18. Haande Haan, L., Ferreira, A.: Extreme Value Theory: An Introduction. Springer (2006)
19. Beirlant, J., Goegebeur, Y., Segers, J., Teugels, J.L.: Statistics of Extremes: Theory and Applications. Wiley (2004)
20. Fraga Alves, I., Neves, C.: Estimation of the finite right endpoint in the Gumbel domain. Statistica Sinica **24**, 1811–1835 (2014)
21. Ferreira, A., de Haan, L.: On the block maxima method in extreme value theory: PWM estimators. Ann. Statist. **43**(1), 276–298 (2015)
22. Pickands, J.: Statistical inference using extreme order statistics. Ann. Stat. 119–131 (1975)
23. Holmes, J., Moriarty, W.: Application of the generalized Pareto distribution to extreme value analysis in wind engineering. J. Wind. Eng. Ind. Aerodyn. **83**(1), 1–10 (1999)
24. Gilli, M., Këllezi, E.: An application of extreme value theory for measuring financial risk. Comput. Econ. **27**(2–3), 207–228 (2006)
25. Falk, M.: Best attainable rate of joint convergence of extremes. In: Extreme Value Theory, pp. 1–9. Springer (1989)
26. Haande Haan, L., Stadtmüller, U.: Generalized regular variation of second order. J. Aust. Math. Soc. **61**(3), 381–395 (1996)

Chapter 4
Extreme Value Statistics

When studying peaks in electricity demand, we may be interested in understanding the risk of a certain large level for demand being exceeded. For example, there is potential interest in finding the probability that the electricity demand of a business or household exceeds the contractual limit. An alternative, yet in principle equivalent way, involves assessment of maximal needs for electricity over a certain period of time, like a day, a week or a season within a year. This would stem from the potential interested in quantifying the largest electricity consumption for a substation, household or business. In either case, we are trying to infer about extreme traits in electricity loads for a certain region assumed fairly homogeneous with the ultimate aim of predicting the likelihood of an extreme event which might have never been observed before. While the exact truth may be not be possible to determine, it may be possible come up with an educated guess (an estimate) and ascertain confidence margins around it.

In this chapter, not only we will list and describe mainstream statistical methodology for drawing inference on extreme and rare events, but we will also endeavour to elucidate what sets the semiparametric approach apart from the classical parametric approach and how these two eventually align with one another. For details on possible approaches and related statistical methodologies we refer to [1, 2].

We hope that, through this chapter, practitioners and users of extreme value theory will be able to see how the theoretical results in Chap. 3 translate in practice and how conditions can be checked. We will be mainly concerned with semi-parametric inference for univariate extremes. This statistical methodology builds strongly on the fundamental results expounded in the previous section, most notably the theory of extended regular variation (see e.g. [3], Appendix B).

Despite the numerous approaches whereby extreme values can be statistically analysed, these are generally classed into two main frameworks: methods for maxima over fixed intervals (blocks) and methods for exceedances (peaks) over high

M. Jacob et al., *Forecasting and Assessing Risk of Individual Electricity Peaks*, Mathematics of Planet Earth,
https://doi.org/10.1007/978-3-030-28669-9_4

thresholds. The former relates the oldest group of models, arising from the seminal work of [4] as block maxima models to be fitted to the largest observations collected from large samples (blocks) of identically distributed observations. The latter has often been considered the most useful framework in practical applications due to the widely proclaimed advantage over the former of a more efficient use of the often scarce extreme data.

The two main settings in univariate extreme value theory we are going to address are:

- The block maxima (BM) method, whereby observations are selected in blocks of equal length (assumed large blocks) and perform inference under the assumption that the maximum in each block (usually a year) is well approximated by the Generalised Extreme Value (GEV) distribution.
- The peaks over threshold method (POT) enables to restrict attention to those observations from the sample that exceed a certain level or threshold, supposedly high, and use mainstream statistical techniques such as maximum likelihood of method of moments to estimation and hypothesis testing, under the assumption that these excesses (values by each the threshold is exceeded) follow exactly a Generalised Pareto distribution (GPD).

Application to energy smart meter data is an important part of the challenge, in the sense of the impeding application to extreme quantile estimation, i.e. to levels which are exceeded with only a small probability. A point to be wary of, when applying EVT is that, due to the wiggliness of the real world and the means by which empirical measurements are collected, observations hardly follow exactly an extreme value distribution. A recent application of extreme value theory can be found in [5]. Despite the underpinning theory to the physical models considered in this paper determines that the true probability function generating the data belongs to the Gumbel domain of attraction, the authors attempt statistical verification of this assumption with concomitant estimation of the GEV shape parameter γ via maximum likelihood, in a purely parametric approach. They find an estimate of -0.002 which indicates that their data is bounded from above, i.e., there is a finite upper endpoint. However this is merely a point estimate whose significance must be evaluated through a test of hypothesis. In the parametric setting, the natural choice would be the likelihood ration test for the simple null hypothesis that $\gamma = 0$.

The semiparametric framework, where inference takes places in the domains of attraction rather than through of the prescribed limiting distribution to the data—either GEV or GPD depending on we set about to look at extremes in the data at our disposal—has proven a fruitful and flexible approach.

In this chapter, we discuss the choice of max-domains of attraction within the semiparametric framework where an EVI $\gamma = 0$ and finite upper endpoint are allowed to coexist. To this effect, we choose to focus on the POT approach as the methodology expounded here will be greatly driven and moderated by the statistical analysis of extreme features (peaks) exhibited by the Irish smart meter data. The description of the data has been given in Chap. 1. We recall that the data comprises 7 weeks of half-hourly loads (in kWh) for 503 households. Figure 4.1 is a rendering of the box-plots

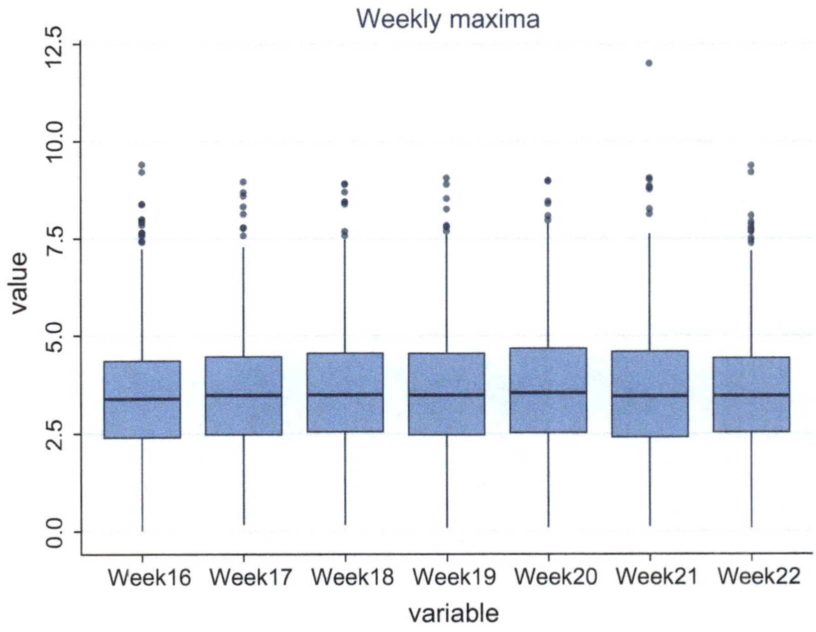

Fig. 4.1 Parallel box-plot representation of weekly maxima

for each week being analysed. All seven box-plots look very similar, both in terms of median energy demand and dispersion of values of energy around the median. There is however one absolute extreme which we will be taking notice in the next sections. This value was recorded in Week 21. An important consideration at this point is that there is a physical limit to the individual electricity loads, which not any less imposed by limitations of the electrical supply infrastructure than by contractual constraint on the upper bound for an individual load. In statistical terms, this means that the assumption of a finite upper endpoint to the d.f. underlying seems fairly reasonable. Nonetheless, the data might indicated otherwise, suggesting that households are actually operating far below the stipulated upper bound, a circumstance that can be potentially exploited by the energy supplier so as to improve management of energy supply.

Figure 4.2 is a scatter-plot representation of the actual observations in terms of the household number. With these two plots we intend to illustrate the two stated methods for the statistical analysis of extreme values, both BM and POT. The top panel shows all the data points. While proceeding with the BM method, one would take the largest observation at each $h = 1, 2, \ldots, 503$, whereby one would have available a sample of 503 independent maxima. On the other hand, applying the POT method with the selected high threshold $t = 7\,\text{kWh}$, we are left with fewer data points and more importantly with several observations originating from the same household. In this case, we find the observations fairly independent only because they are one week apart. We highlight that the POT method implies that some households are naturally discarded, which we find an important caveat to the POT-method, a method that has heralding the efficient use the available extreme data.

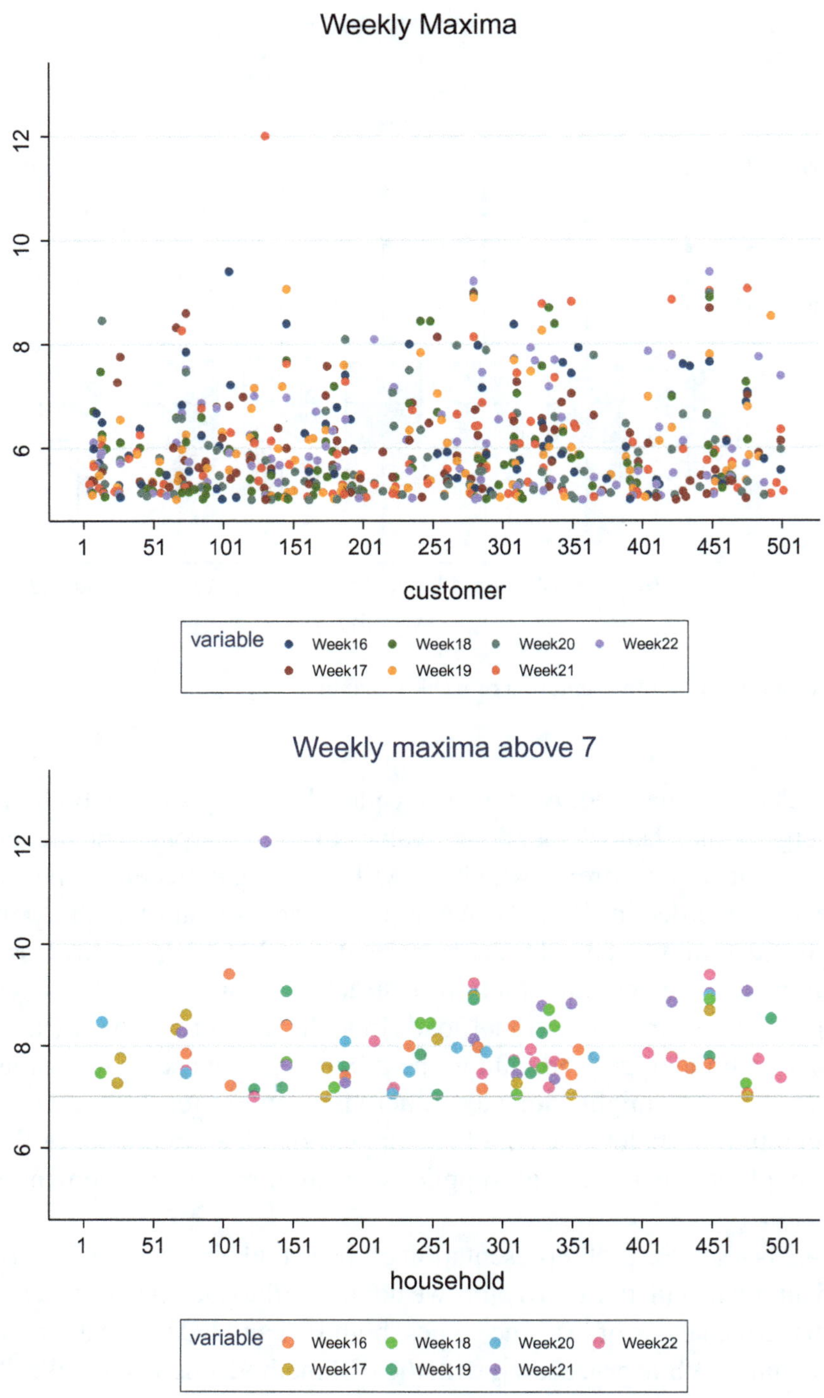

Fig. 4.2 Weekly maxima and exceedances above the threshold 7 kWh for the Iris smart meter data

4.1 Block Maxima and Peaks over Threshold Methods

Due to their nature, semi-parametric models, are never specified in detail by hand. Instead, the only assumption made is that F is in the domain of attraction of an extreme value distribution, i.e. $F \in \mathcal{D}(G_\gamma)$. In order to better understand what we mean by inference in extreme domains of attraction, let us remind ourselves of the well-known condition of extended regular variation [3, 6, 7], introduced in Chap. 3, as tantamount to the domain of attraction condition. Notably, $F \in \mathcal{D}(G_\gamma)$ if and only if there exists a positive measurable function a such that the pertaining tail quantile function $U \in ERV_\gamma$. The limit in (3.10) coincides with the U-function of the GPD, with distribution function $1 + \log G_\gamma$, which justifies the usual inference on the excesses above a high threshold ascribed to the POT method. In fact, the extreme value condition (3.10) on the tail quantile function U is the usual assumption in semi-parametric inference for extreme outcomes. We shall develop this aspect further in Sect. 4.3. In the next Sect. 4.2 we will start off with yet another equivalent extreme value condition to the extended regular variation of U that is provided in [8] for dealing with block length and/or block number as opposed to looking at a certain of upper order statistics above a sufficiently high (random) threshold. Let V be the left generalised inverse of $-1/\log F$, i.e. $V\big(-1/\log(1-t)\big) = F^{\leftarrow}(1-t)$. In other words, $V(t) = F^{\leftarrow}(e^{-1/t})$, for $0 \le t < 1$, which conveys standardisation to the standard Fréchet. It is straightforward to see that the df F underlying the random sample $(X_1, \ldots, X_n)$ belongs to some max-domain of attraction if and only if there exist functions a and b, as defined in Theorem 3.2, such that

$$\lim_{t\to\infty} \frac{V(tx) - b(t)}{a(t)} = \frac{x^\gamma - 1}{\gamma}, \tag{4.1}$$

for all $x > 0$. In contrast to the previous case of associating relation (3.1) with (3.10), there is now an asymptotically negligible factor creeping in when substituting $b(t)$ with $V(t)$. This is where we focus next, as this factor reflects the bias stemming from absorbing b (or V) into the location parameter of the GEV limit distribution (see 3.2). The common understanding is that such bias is somewhat difficult to control, but we will have a closer look at this in terms of the second order refinements. The theoretical development for working out the order of convergence in Eq. (4.1) and (3.10) in tandem is given in Proposition 4.1. For the proof, we refer the reader to [9].

Proposition 4.1 *Assume condition (3.10) (i.e. $F \in \mathcal{D}(G_\gamma)$) and that U is of extended regular variation of second order, that is, there exists a positive or negative function A with $\lim_{t\to\infty} A(t) = 0$ and a non-positive parameter ρ, such that for $x > 0$,*

$$\lim_{t\to\infty} \frac{\frac{U(tx)-U(t)}{a(t)} - \frac{x^\gamma - 1}{\gamma}}{A(t)} = \frac{1}{\rho}\Big(\frac{x^{\gamma+\rho}-1}{\gamma+\rho} - \frac{x^\gamma - 1}{\gamma}\Big) =: H_{\gamma,\rho}(x). \tag{4.2}$$

Define

$$\widetilde{A}(t) := \begin{cases} \frac{1-\gamma}{2}t^{-1}, & \gamma \neq 1,\ \rho < -1, \\ A(t) + \frac{1-\gamma}{2}t^{-1}, & \gamma \neq 1,\ \rho = -1, \\ A(t), & \rho > -1 \text{ or } (\gamma = 1,\ \rho > -2), \\ A(t) + \frac{1}{12}t^{-2}, & \gamma = 1,\ \rho = -2, \\ \frac{1}{12}t^{-2}, & \gamma = 1,\ \rho < -2. \end{cases}$$

If $\tilde{A}$ is either a positive or negative function near infinity, then with

$$\tilde{a}(t) := \begin{cases} a(t)\left(1 + \frac{\gamma-1}{2}t^{-1}\right), & \gamma \neq 1,\ \rho \leq -1, \\ a(t), & \rho > -1 \text{ or } (\gamma = 1,\ \rho > -2), \\ a(t)\left(1 - \frac{1}{12}t^{-2}\right), & \gamma = 1,\ \rho \leq -2, \end{cases}$$

the following second order condition holds

$$\lim_{t\to\infty} \frac{\frac{V(tx)-V(t)}{\tilde{a}(t)} - \frac{x^{\gamma}-1}{\gamma}}{\widetilde{A}(t)} = H_{\gamma,\tilde{\rho}}(x), \tag{4.3}$$

for all $x > 0$, where $\tilde{\rho} = \max(\rho, -1)$ if $\gamma \neq 1$, and $\tilde{\rho} = \max(\rho, -2)$ if $\gamma = 1$.

We now provide four examples of application of Proposition 4.1 alongside further details as to how the prominent GPD can, at a first glance, escape the grasp of this proposition.

Example 4.1

Burr$(1, \tau, \lambda)$. This example develops along similar lines to the proof of Proposition 4.1. The Burr distribution, with d.f. $1 - (1 + x^{\tau})^{-\lambda}$, $x \geq 0$, $\lambda, \tau > 0$, provides a very flexible model which mirrors well the GEV behaviour in the limit of linearly normalised maxima, also allowing a wide scope for tweaking the order of convergence through changes in the parameter λ. The associated tail quantile function is $U(t) = (t^{1/\lambda} - 1)^{1/\tau}$, $t \geq 1$. Upon Taylor's expansion of U, the extreme tail condition up to second order (Eq. 4.2) arises:

$$U(tx) - U(t) = \frac{t^{\frac{1}{\lambda\tau}}}{\lambda\tau}\left[\frac{x^{\frac{1}{\lambda\tau}} - 1}{\frac{1}{\lambda\tau}} - \lambda t^{-1/\lambda}\left(x^{\frac{1}{\lambda}(\frac{1}{\tau}-1)} - 1\right) + o\left(t^{-1/\lambda}\right)\right],$$

as $t \to \infty$. Whence, the second order condition on the tail given in Eq. (4.2) holds for $\gamma = 1/(\lambda\tau)$ and $\rho = -1/\lambda$, $\gamma + \rho \neq 0$, with

$$a(t) = \frac{t^{\frac{1}{\lambda\tau}}}{\lambda\tau}\left(1 - \left(\frac{1}{\tau} - 1\right)t^{-\frac{1}{\lambda}}\right) \text{ and } A(t) = \frac{1}{\lambda}\left(\frac{1}{\tau} - 1\right)t^{-\frac{1}{\lambda}} = (\gamma + \rho)t^{\rho}.$$

Proposition 4.1 is clearly applicable and therefore the Burr distribution satisfies the extreme value condition of second order (Eq. 4.3) with $\gamma = 1/(\gamma\tau)$ and $\tilde{\rho} = \max(-1/\lambda, -1)$ if $\tau \neq 1$.

Example 4.2

Cauchy. The relevant d.f. is $F(x) = \frac{1}{\pi}\arctan x + \frac{1}{2}$, $x \in$ The corresponding tail quantile function is $U(t) = \tan\big(\pi/2 - \pi/t\big) = t/\pi - \pi/3\,t^{-1} + O(t^{-3})$, as $t \to \infty$, and admits the representation $U(tx) - U(t) = \frac{t}{\pi}\big[x - 1 - \frac{\pi^2}{3}t^{-2}(x^{-1} - 1) + O(t^{-4})\big]$, $x > 0$. Hence, we have that $\gamma = 1$, $\rho = -2$ in Eq. (4.2) with auxiliary function $a(t) = t/\pi$. Proposition (4.1) thus ascertains that (Eq. 4.3) also holds true for the Cauchy distribution where $\gamma = 1$ and $\tilde{\rho} = -2$.

Example 4.3

GPD(γ). The relevant d.f. is $W_\gamma(x) = 1 - (1 + \gamma x)^{-1/\gamma}$, for all x such that $1 + \gamma > 0$. The pertaining tail quantile function is $U(t) = (t^\gamma - 1)/\gamma$ which is also born out of the exact tail condition (3.10). Clearly, U does not satisfy the second order condition (Eq. 4.2) in a straightforward fashion, however we are going to show that the corresponding $V(t) = U\big(1/(1 - e^{-1/t})\big)$ satisfies (Eq. 4.3). To this end, we shall deal with the cases $\gamma = 1$ and $\gamma \neq 1$ separately.

Case $\gamma = 1$: Applying Laurent series expansion upon $(1 - e^{-1/t})^{-1}$, we get

$$V(tx) - V(t) = \Big(1 - \frac{1}{12t}\Big)(x - 1) + \frac{2}{12t}H_{1,-2}(x) + O(t^{-3}),$$

as $t \to \infty$. Whence, the second order condition (Eq. 4.3) holds with $\gamma = 1$ and $\tilde{\rho} = -2$, where $\widetilde{A}(t) = t^{-2}/6$ and $\widetilde{a}(t) = t(1 + \widetilde{A}(t)/\tilde{\rho})$.
Case $\gamma \neq 1$: Upon Taylor's expansion around zero, we obtain

$$V(tx) - V(t) = t^\gamma\bigg[\Big(1 + \frac{\gamma - 1}{2t}\Big)\frac{x^\gamma - 1}{\gamma} - \frac{\gamma - 1}{2t}H_{\gamma,-1}(x)\bigg] + O(t^{-3}),$$

as $t \to \infty$. Whence, the second order condition (Eq. 4.3) holds with $\rho = -1$, where $t\widetilde{A}(t) = (1 - \gamma)/2$ and $\widetilde{a}(t) = t^\gamma(1 + \widetilde{A}(t)/\tilde{\rho})$.

Therefore, the GPD verifies Proposition 4.1 if one tunnels through the consideration that the GDP satisfies (Eq. 4.2) with $\rho = -\infty$.

Example 4.4

Pareto(α). This distribution is a particular case of the GPD d.f. in Example 4.1 with $\gamma = 1/\alpha > 0$ and $U(t) = t^{1/\alpha}$, that is U does not satisfy the second order condition (Eq. 4.2) and thus Proposition 4.1 stands applicable provided similar interpretation to Example 4.1.

Example 4.5

Contaminated Pareto(α). We now consider the Pareto distribution with a light contamination in the tail by a slowly varying function $L(t) = (1 + \log t)$, that is, $L(tx)/L(t) \to 1$, as $t \to \infty$, for all $x > 0$. This gives rises to the quantile function $U(t) = t^{1/\alpha}(1 + \log t)$, with $\alpha > 0$. For the sake of simplicity, we shall use the identification $\gamma = 1/\alpha$. With some rearrangement, we can write the spacing $U(tx) - U(t)$ in such a way that the first and second order parameters in condition (Eq. 4.2), both γ and $\rho \leq 0$, crops up: $U(tx) - U(t) = \gamma t^{\gamma}(\log t + 1)\big[\big(1 + \frac{1}{\gamma \log t + 1}\big)\frac{x^{\gamma}-1}{\gamma} + \frac{1}{1+\log t} H_{\gamma,0}(x)\big]$, where $H_{\gamma,0}(x) := \frac{1}{\gamma}\big(x^{\gamma}\log x - \frac{x^{\gamma}-1}{\gamma}\big)$. Note that we have provided an exact equality, i.e. there is no error term. We thus find that tampering with the Pareto distribution, by contaminating its tail-related values with a slowly varying factor, is just enough bring the convergence (Eq. 4.2) to a halt which is flagged-up by the lowest possible $\rho = 0$. This stalling of the Pareto distribution enables to fullfil the conditions in Proposition (4.1) thus ensuring that this contaminated Pareto distribution belongs to the max-domain of attraction of the GEV distribution with $\gamma = 1/\alpha > 0$ and $\tilde{\rho} = 0$.

4.2 Maximum Lq-Likelihood Estimation with the BM Method

Let us define the random sample consisting of k i.i.d. block maxima as

$$M_i = \max_{(i-1)m<j\leq im} X_j, \qquad i = 1, 2, \ldots, k, \; m = 1, 2, \ldots \tag{4.4}$$

The above essentially states that we are dividing the whole sample of size n into k blocks of equal length (time) m. For the Extreme Value theorem to hold within each block, the block length must be sufficiently large, i.e. one needs to impose m tending to infinity to able to proceed with inference. We are then led to the reasonable assumption that the sample of k-maxima behaves approximately as though it stems from the GEV.

Under a semi-parametric approach, maximum likelihood estimators for the vector-valued parameter $\theta = (\mu, \sigma, \gamma)$ are obtained by pretending (which is approximately true) that the random variables $M_1, M_2, \ldots, M_k$ are independent and identically distributed as maxima of the GEV distribution with d.f. given by

$$G_\theta(x) = \exp\Big\{-\Big(1+\gamma\frac{x-\mu}{\sigma}\Big)^{-1/\gamma}\Big\},$$

for those x such that $\sigma + \gamma(x-\mu) > 0$. The density of the parametric fit to the BM framework is the GEV density, which we denote by g_θ, may be differ slightly from the true unknown p.d.f. f underlying the sampled data. We typically estimate these constants $a(m)$ and $b(m)$ via maximum likelihood, despite these being absorbed into the scale $\sigma > 0$ and location $\mu \in$ parameters of the parametric limiting distribution thus assumed fixed, eventually. As a result, BM-type estimators are not so accurate for small block sizes since these estimators must rely on blocks of reasonable length to fulfill the extreme value theorem.

The criterion function ℓ_q that gives rise to a maximum Lq-likelihood (MLq) estimator,

$$\widetilde{\theta} := \arg\max_{\theta\in\Theta} \sum_{i=1}^{k} \ell_q(g_\theta(x_i)),$$

for $q \geq 0$, makes use of the Tsallis deformed logarithm as follows:

$$\widetilde{\theta} = \arg\max_{\theta\in\Theta} \sum_{i=1}^{k} \frac{\big(g_\theta(x_i)\big)^{1-q} - 1}{1-q}, \quad q \geq 0, \tag{4.5}$$

This MLq estimation method picks up the standard maximum likelihood estimator (MLE) if one sets the distortion parameter $q = 1$. This line of reasoning can be stretched on to a continuous path, that is, as q tends to 1, the MLq estimator approaches the usual MLE. The common understanding is that values of q closer to one are preferable when we have numerous maxima drawn from large blocks since this will give enough scope for the EVT to be accessible and applicable. In practice, we often encounter limited sample sizes $n = m \times k$ in the sense that either a small number of extremes (k sample maxima) or blocks of insufficient length m to contain even one extreme are available. MLq estimators have been recognised as particularly useful in dealing with small sample sizes, which is often the situation in the context of the analysis of extreme values due to the inherent scarcity of extreme events with catastrophic impact. Previous research by [10, 11] shows that the main contribution towards the relative decrease in the mean squared error stems from the variance reduction, which is the operative statement in small sample estimation. This is in contrast with the bias reduction often sought after in connection with large sample inference. Large enough samples tend to yield stable and smooth trajectories in the estimates-paths, allowing scope for bias to set in, and eventually reflecting the regularity conditions in the maximum likelihood sense. Once these asymptotic conditions are attained, the dominant component of the bias starts to emerge, and by then it can be efficiently removed. This is often implemented at the expense of

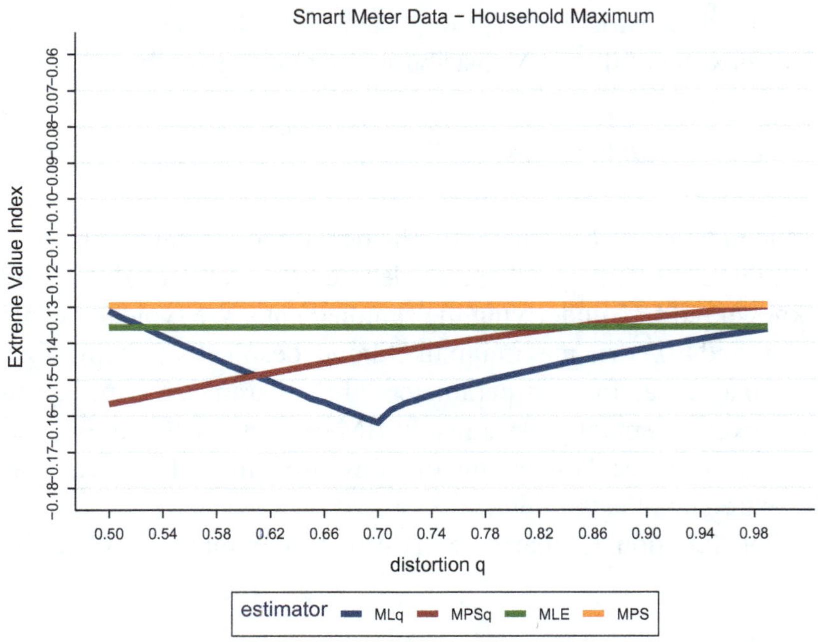

Fig. 4.3 Estimation of the EVI with the BM method

an increased variance, for the usual bias/variance trade off in statistics seems never to offer anything worthwhile on one side without also providing a detriment to the other.

Furthermore, we will address how the maximum likelihood compares with the maximum product of spacings (MPS) estimator in this case study. The MPS estimator of θ maximises the product of spacings

$$\prod_{i=1}^{k+1} D_i(\theta) = \prod_{i=1}^{k+1} \Big\{ G_\theta(x_{i,k}) - G_\theta(x_{i-1,k}) \Big\},$$

with $G_\theta(x_{0,k}) = 0$ and $G_\theta(x_{k+1,k}) = 1$, or equivalently the log-spacings

$$L^{MPS}(\theta; \mathbf{x}) = \sum_{i=1}^{k+1} \log D_i(\theta). \tag{4.6}$$

The MPS method was introduced by [12], and independently by [13]. A generalisation of this method is proposed and studied in great depth by [14]. The MPS method was further exploited by [15] in estimating and testing for the only possible three types of extreme value distributions (Fréchet, Gumbel and Weibull), all unified in the GEV distribution.

Figure 4.3 displays the sample paths of the adopted extreme value index estimators, plotted against several values of the distortion parameter $q \in [0.5, 1]$. As q increases to 1, the deformed version of both ML and MPS estimators approach their classical counterpart which stem from stipulating the natural logarithm as criterion function. The estimates for the EVI seem to consolidate between −0.14 and −0.13. The negative estimates obtained for all values of q provide evidence that the true d.f. F generating the data belongs to the Weibull domain of attraction and therefore we can reasonable conclude that we are in the presence of a short tail with finite right endpoint. The next section concerns the estimation of this *ultimate return-level*.

4.2.1 Upper Endpoint Estimation

A class of estimators for the upper endpoint x^F stems from the extreme value condition (Eq. 4.1) via the approximation $V(\infty) \approx V(m) - a_m/\gamma$, as $m \to \infty$, by noticing that $V(\infty) = \lim_{t\to\infty} V(t) = F^{\leftarrow}(1) = x^F$. The existing finite right endpoint x^F can be viewed as the ultimate return level. When estimating extreme characteristics of this sort, we are required to replace all the *unknowns* in the above by their empirical analogues, yielding the estimator for the right endpoint:

$$\hat{x}^F := \hat{V}(m) - \frac{\hat{a}(m)}{\hat{\gamma}}, \tag{4.7}$$

where quantities $\hat{a}$, $\hat{V}$ and $\hat{\gamma}$ stand for the MLq estimators for the scale and location functions $a(m)$ and $V(m)$, and for the EVI, respectively.

Figure 4.4 displays the endpoint estimates for several values of $q \leq 1$ with respect to the tilted version of both ML and MPS estimators. The latter always finds larger estimates than the former, with a stark distance from the observed overall maximum.

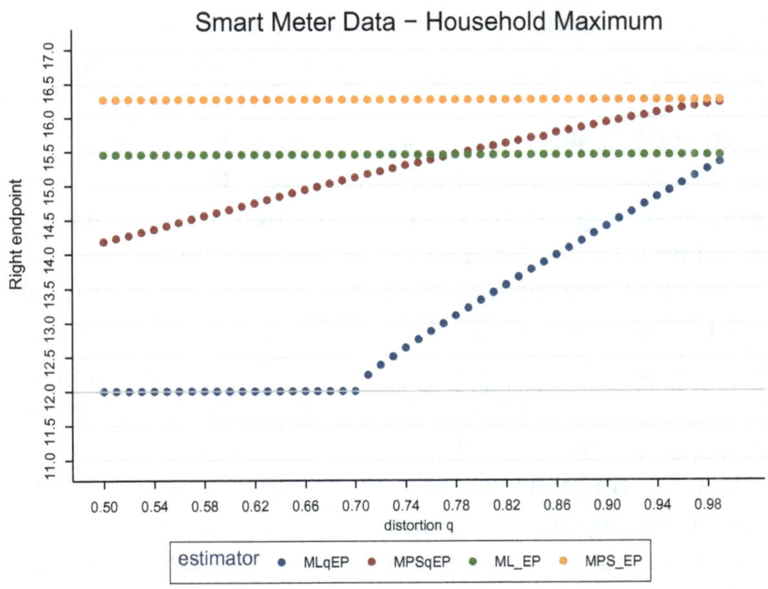

Fig. 4.4 Estimation of the upper endpoint through the BM method

Since the adopted estimators do not seem to herd towards one value, it is not easy to conciliate between them. Given the maximum likelihood estimator has been widely used and extensively studied in the literature, it is sensible to ascribe preference to this estimator. Furthermore, since we are dealing with small sample sizes (we are taking the maximum over 7 weeks), the distorted version, i.e., the MLq estimator must be taken into account. Therefore, we find reasonable to take as estimate for the upper bound the midpoint of the region where the MLq changes way and travels across towards the plain ML estimator with increasing q. Thus, we get an upper bound of 14.0 kWh, approximately.

4.3 Estimating and Testing with the POT Method

In Sect. 4.1, we noticed that the function appearing in the limit of the extended regular variation of U matches the tail quantile function of the Generalised Pareto distribution. This fact reflects indeed the exceptional role of the GPD in the extreme value theory for exceedances [16, 17] and prompts the need for classifying of the tails of all possible distributions in $\mathcal{D}(G_\gamma)$ into three classes in accordance to the sign of the extreme value index γ. For positive γ, the power-law behaviour of the underlying tail distribution function $1 - F$ has important implications one of which being the presence of infinite moments. Because when $\gamma > 0$ the first order condition (3.10) can be rephrased as $\lim_{t\to\infty} U(tx)/U(t) = x^\gamma$, for all $x > 0$, that is, U is γ-regularly varying at infinity (notation: $U \in RV_\gamma$), then Karamata's theorem ascertains that $E(X_1^+)^p$ is infinite for $p > 1/\gamma$, where $X_1^+ = \max(0, X_1)$. Hence, heavy-tailed distributions in a max-domain of attraction not only have an infinite right endpoint, but also the order of finite moments is determined by order of the magnitude of the EVI $\gamma > 0$. The Fréchet domain of attraction contains distributions with polynomially decay tails such as the Pareto, Cauchy, Student's and Fréchet itself. All d.f.'s belonging to $\mathcal{D}(G_\gamma)$ with $\gamma < 0$—Weibull domain of attraction—are light tailed distributions with finite right endpoint. Such domain of attraction encloses Uniform and Beta distributions. The intermediate case $\gamma = 0$ is of particular interest in many applied sciences where extremes are relevant. At a first glance, the Gumbel domain of attraction seems quite appealing for the simplicity of inference in connection to G_0 that dispenses the estimation of γ. But a closer inspection allows to appreciate the great variety of distributions possessing such an exponential tail related behaviour, whether having finite upper endpoint or not. Normal, Gamma and Lognormal distributions can all be found in Gumbel domain. The negative Fréchet distribution also belongs to the Gumbel domain, albeit with finite endpoint (cf. [18]). Therefore, a statistical test for assessing significance of the EVI would be of great use and most consequential. Looking for the most propitious type of tail before estimating tail-related features of the distribution underlying the data can mark the difference between aiming at the estimation of extreme quantile or sprinting to the estimation of the upper endpoint estimation. In fact, adopting tailored statistical methodology to the suitable domain of attraction for the underlying d.f. F has become a regular practice.

4.3.1 Selection of the Max-Domain of Attraction

A test for Gumbel domain *versus* Fréchet or Weibul max-domains has received in the literature the general designation of statistical choice of extreme domains of attraction. References in this respect are [19–29].

The present section primarily deals with the two-sided problem of testing Gumbel domain against Fréchet or Weibull domains, i.e.,

$$F \in \mathcal{D}(G_0) \quad vs \quad F \in \mathcal{D}(G_\gamma)_{\gamma \neq 0}. \tag{4.8}$$

Bearing on the sample maximum as the dominant and most informative statistic in any analysis of extreme value, we shall consider the ratio statistic

$$T_n^*(k) := T_n(k) - \log k = \frac{X_{n,n} - X_{n-k,n}}{\frac{1}{k}\sum_{i=1}^{k}\left(X_{n-i+1,n} - X_{n-k,n}\right)} - \log k, \tag{4.9}$$

for the testing problem in Eq. (4.8). According to the asymptotic results stated in [27], the null hypothesis $F \in \mathcal{D}(G_0)$ is rejected, against the bilateral alternative $F \in \mathcal{D}(G_\gamma)_{\gamma \neq 0}$, at an asymptotic level $\alpha \in (0, 1)$ if $T_n^*(k) < g_{\alpha/2}$ or $T_n^*(k) > g_{1-\alpha/2}$, where g_ε denotes the ε-quantile of the Gumbel distribution, i.e., $g_\varepsilon = -\log(-\log \varepsilon)$. One sided tests are also within reach of this test statistic. We reject the null hypothesis in favor of either unilateral alternatives $H_1^{'} : F \in \mathcal{D}(G_\gamma)_{\gamma<0}$ or $H_1^{''} : F \in \mathcal{D}(G_\gamma)_{\gamma>0}$ on either side $T_n^*(k) < g_\alpha$ or $T_n^*(k) > g_{1-\alpha}$, respectively. Neves et al. [27] show that this statistic provides a consistent test to discriminate between light tails and heavy tails, departing from the null hypothesis of an exponential tail.

Built on Shapiro–Wilk goodness-of-fit statistic, the well-known Hasofer and Wang test statistic, embodies the reciprocal squared empirical coefficient of variation associated with the sample of the excesses above the random threshold $X_{n-k,n}$. More concretely,

$$W_n(k) := \frac{1}{k}\frac{\left(k^{-1}\sum_{i=1}^{k} Z_i\right)^2}{k^{-1}\sum_{i=1}^{k} Z_i^2 - \left(k^{-1}\sum_{i=1}^{k} Z_i\right)^2}, \tag{4.10}$$

where $Z_i := X_{n-i+1,n} - X_{n-k,n}$, $i = 1, \ldots, k$. Giving heed to the problem of the statistical choice of domains of attraction postulated in Eq. (4.8), we define for $j = 1, 2$

$$N_{n,k}^{(j)} := \frac{1}{k}\sum_{i=1}^{k}(X_{n-i+1,n} - X_{n-k,n})^j \tag{4.11}$$

and use it to write Hasofer and Wang, $W_n(k)$, and Greenwood $R_n(k)$ statistics in the following way:

$$R_n(k) = \frac{N_n^{(2)}(k)}{\left(N_n^{(1)}(k)\right)^2}, \tag{4.12}$$

$$W_n(k) = \frac{1}{k}\left[1 - \frac{R_n(k) - 2}{1 + (R_n(k) - 2)}\right]. \tag{4.13}$$

Considering, as before, the k upper order statistics from a sample of size n such that $k = k_n$ is an intermediate sequence, i.e., $k \to \infty$ and $k/n \to 0$ as $n \to \infty$, define

$$R_n^*(k) := \sqrt{k/4}\,\big(R_n(k) - 2\big) \tag{4.14}$$

$$W_n^*(k) := \sqrt{k/4}\,\big(kW_n(k) - 1\big). \tag{4.15}$$

These normalized versions, $R_n^*(k)$ and $W_n^*(k)$, are eventually the main features to take part in the testing procedure. The critical region for the two-sided test of nominal size α is given by $|T_n^*(k)| > z_{1-\alpha/2}$, with z_ε denoting the ε-quantile of the standard normal distribution and where T has to be conveniently replaced by R or W.

In addition, one-sided testing problems

$$F \in \mathcal{D}(G_0) \quad vs \quad F \in \mathcal{D}(G_\gamma)_{\gamma<0} \quad \big(\text{or } F \in \mathcal{D}(G_\gamma)_{\gamma>0}\big),$$

can also be tackled with both these test statistics. Here, the null hypothesis is rejected in favor of either unilateral alternatives $H_1^{'} : F \in \mathcal{D}(G_\gamma)_{\gamma<0}$ or $H_1^{''} : F \in \mathcal{D}(G_\gamma)_{\gamma>0}$ on either side $T_n^*(k) < z_\alpha$ or $T_n^*(k) > z_{1-\alpha}$, respectively, with z_ε denoting again the ε-quantile of the standard normal distribution and whereas T has to be conveniently replaced by R or W.

We remark that the test based on the very simple Greenwood-type statistic R^* is shown to good advantage when testing the presence of heavy-tailed distributions is in demand. While the R^*-based test barely detects small negative values of γ, the Hasofer and Wang's is the most powerful test under study concerning alternatives in the Weibull domain of attraction. The three testing procedures will be illustrated with the Iris smart meter data in the next section.

4.3.2 Testing for a Finite Upper Endpoint

The aim of this section is to assess finiteness in the right endpoint of the actual d.f. F underlying the Irish smart meter data. The basic assumption is that F belongs to some max-domain of attraction. We then consider the usual asymptotic setting, where assume an intermediate number k of order statistics to drawn inference upon, that is, we take $k = k_n \to \infty$ and $k_n/n \to 0$, as $n \to \infty$, and hence the corresponding random threshold $X_{n-k,n} \to x^F$ *a.s.*.

The statistical judgment on whether a finite upper bound exists finite will be informed by the testing procedure introduced in [30]. Notably, the testing problem

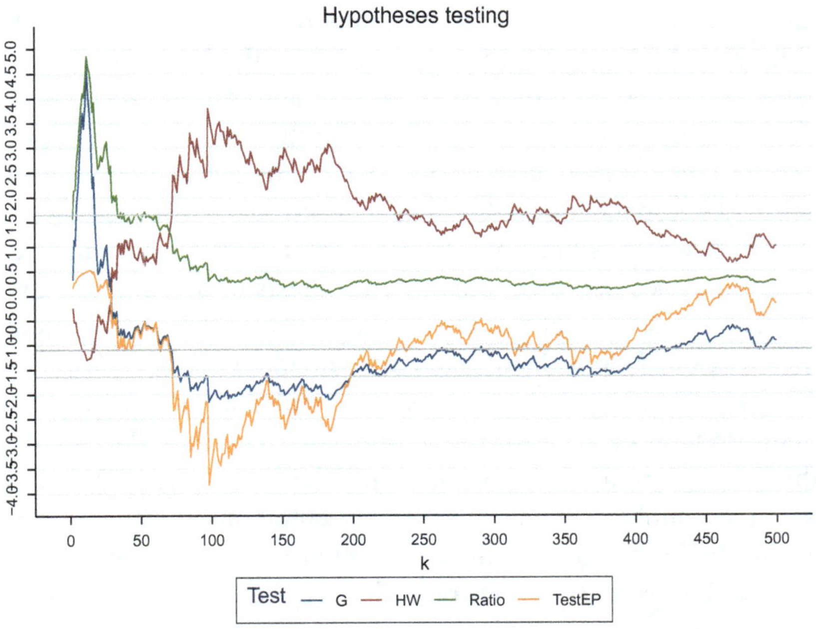

Fig. 4.5 Statistical choice of max-domain of attraction and detection of finite endpoint

$$H_0 : F \in \mathcal{D}(G_0)\,,\ x^F = \infty \quad vs \quad H_1 : F \in \mathcal{D}(G_\gamma)_{\gamma \le 0}\,,\ x^F < \infty$$

can be tackled using the log-moments

$$M_{n,k}^{(j)} := \frac{1}{k}\sum_{i=0}^{k-1}\left(\log X_{n-i,n} - \log X_{n-k,n}\right)^j, \quad j = 1, 2. \tag{4.16}$$

The test statistic T_1 being used is defined as

$$T_1 := \frac{1}{k}\sum_{i=1}^{k}\frac{X_{n-i,n} - X_{n-k,n} - T}{X_{n,n} - X_{n-k,n}}, \quad \text{with } T := X_{n-k,n}\frac{M_1}{2}\left(1 - \frac{[M_1]^2}{M_2}\right)^{-1}.$$

Under H_0, the standardised version of the test, $T_1^* := \sqrt{k}\,\log k\, T_1$, is asymptotically normal. Moreover, T_1^* tends to inflect to the left for bounded tails in the Weibull domain and to the right if the underlying distribution belongs to the Gumbel domain. The rejection region of the test is given by $|T_1^*| \ge z_{1-\alpha/2}$, for an approximate α significance level. Figure 4.5 displays the sample path of T_1^*, labeled TestEP, alongside the observed values of the three test for selecting max domain of attraction presented in Sect. 4.3.1. The horizontal grey lines mark the critical barriers for the one-sided test at a $\alpha = 5\%$ significance level. When these critical bounds are crossed, the null hypothesis of the Gumbel domain is rejected in favour of the Weibull domain. The statement for the testing problem on the upper endpoint is slightly different as it entails a different breakdown of the Gumbel domain. The choice of the optimal number k of intermediate order statistics is of paramount importance to any inference problem in extremes. Many methods have been proposed, but sadly there is not universal solution that can hold for the multitude of estimators and testing procedures available. Here, we loosely follow the guideline of [23] in that the most adequate choice of the intermediate number k (which carries over to the subsequent semi-parametric inference) should set on the lowest k at which the critical barriers are overtaken. The ratio test (Eq. 4.9), which is known to be the most conservative of all three tests for choice of domains, does not reject the null hypothesis of the Gumbel domain since the green trajectory remains above the second horizontal line from below, for all intermediate values of k considered. We have remarked that the Hasofer and Wang (HW) test defined in Eq. (4.13) is the most powerful test for detecting distributions in the Weibull domain of attraction. The application of the HW test seems to do justice to this Irish smart meter data set and finds sufficient evidence to reject the null hypothesis of the Gumbel domain in favour of entertaining estimation procedures suited to the Weibull domain. Therefore we will proceed on to the estimation of the finite upper bound in the POT framework. This will be tackled in the next section.

4.3.3 Upper Endpoint Estimation

Along similar lines to Sect. 4.2, a valid estimator for the upper endpoint $x^F = U(\infty)$ arises by making $t = n/k$ in the approximate equality corresponding to (3.10), and then replacing $U(n/k)$, $a(n/k)$ and γ by suitable consistent estimators, i.e.

$$\hat{x}^F = \hat{U}\left(\frac{n}{k}\right) - \frac{\hat{a}\left(\frac{n}{k}\right)}{\hat{\gamma}}$$

(cf. Sect. 4.5 of [3]). Typically we consider the semiparametric class of endpoint estimators as follows:

$$\hat{x}^F = X_{n-k,n} - \frac{\hat{a}(n/k)}{\hat{\gamma}}. \tag{4.17}$$

For the application of EVT to the class (Eq. 4.17) of upper endpoint estimators it is thus necessary to estimate the parameter γ. We mention the estimators: Pickands' estimator [17] for $\gamma \in \mathbb{R}$; the so-called ML estimator [31], valid for $\gamma > -1/2$; the moment estimator [32] and the mixed moment estimator, both valid for any $\gamma \in \mathbb{R}$. These moment estimators are purely semiparametric estimators for they are devised upon conditions of regular variation underpinning the max-domain of attraction characterisation. Since these are rather qualitative conditions, with functions U and a specific to the underlying d.f. F, then inference must be built on summary statistics that can capture the most distinctive traits in tail-related observations. The method of moments is ideally suited to this purpose.

In order to develop a novel estimator for the extreme value index $\gamma \in \mathbb{R}$, [33] considered a combination of Theorems 2.6.1 and 2.6.2 of [7] and went on with replacing

F with its empirical counterpart F_n and t by the order statistic $X_{n-k,n}$ with $k < n$. This led to the statistic

$$\hat{\varphi}_n(k) := \frac{M_n^{(1)}(k) - L_n^{(1)}(k)}{\left(L_n^{(1)}(k)\right)^2}, \tag{4.18}$$

where we define, for $j = 1, 2$,

$$L_n^{(j)}(k) := \frac{1}{k}\sum_{i=0}^{k-1}\left(1 - \frac{X_{n-k,n}}{X_{n-i,n}}\right)^j \tag{4.19}$$

and with $M_n^{(j)}(k)$ given in Eq. (4.16). The statistic in (Eq. 4.18) is easily transformed into the so-called mixed moment estimator (MM) for the extreme value index $\gamma \in \mathbb{R}$:

$$\hat{\gamma}_n^{MM}(k) := \frac{\hat{\varphi}_n(k) - 1}{1 + 2\,\min\left(\hat{\varphi}_n(k) - 1, 0\right)}. \tag{4.20}$$

The most attractive features of this estimator are:

- like Pickands and Moment estimators, it is valid for any $\gamma \in \mathbb{R}$ and, contrary to the maximum likelihood estimator, it has a simple explicit functional form;
- it is very close to the maximum likelihood estimator for $\gamma \geq 0$;
- If $\gamma \leq 0$, the asymptotic variance is the same as of the Moment estimator; if $\gamma > 0$, its asymptotic variance is equal to that of the maximum likelihood estimator;
- A shift invariant version with similar properties is available with the same asymptotic variance without sacrificing the dominant component of the bias, which is never increased as long as we keep a suitable number k;
- there are accompanying shift and scale estimators that make e.g. high quantile and upper endpoint estimation straightforward.

Figure 4.6 shows the sample paths of several estimators of the EVI as the upper number k of order statistics embedded in the estimators increases, concomitantly lowering the threshold until the value 5 khW is reached. The standard practice for drawing meaningful conclusions from this type of plots is by eyeballing the trajectories and seek for a plateau of stability at the confluence of the adopted estimators. In the top panel of Fig. 4.6, the MLq estimator of the extreme value index γ, which has no explicit closed form and thus delivers estimates numerically, experiences convergence issues. This is often the case with maximum likelihood estimation for the GPD when the true value is negative but close to zero.

In the semi-parametric setting, whilst working in the domain of attraction rather than dealing with the limiting distribution itself, the upper intermediate order statistic $X_{n-k,n}$ plays the role of the high deterministic threshold $u \uparrow x^F \leq \infty$ above which the parametric fit to the GPD is applicable. For the asymptotic properties of the POT maximum likelihood estimator of the EVI under a semi-parametric approach, see e.g. [34–36]. Although theoretically well determined, even when $\gamma \uparrow 0$, the non-convergence to a ML-solution can be an issue when γ is close to zero. There are also

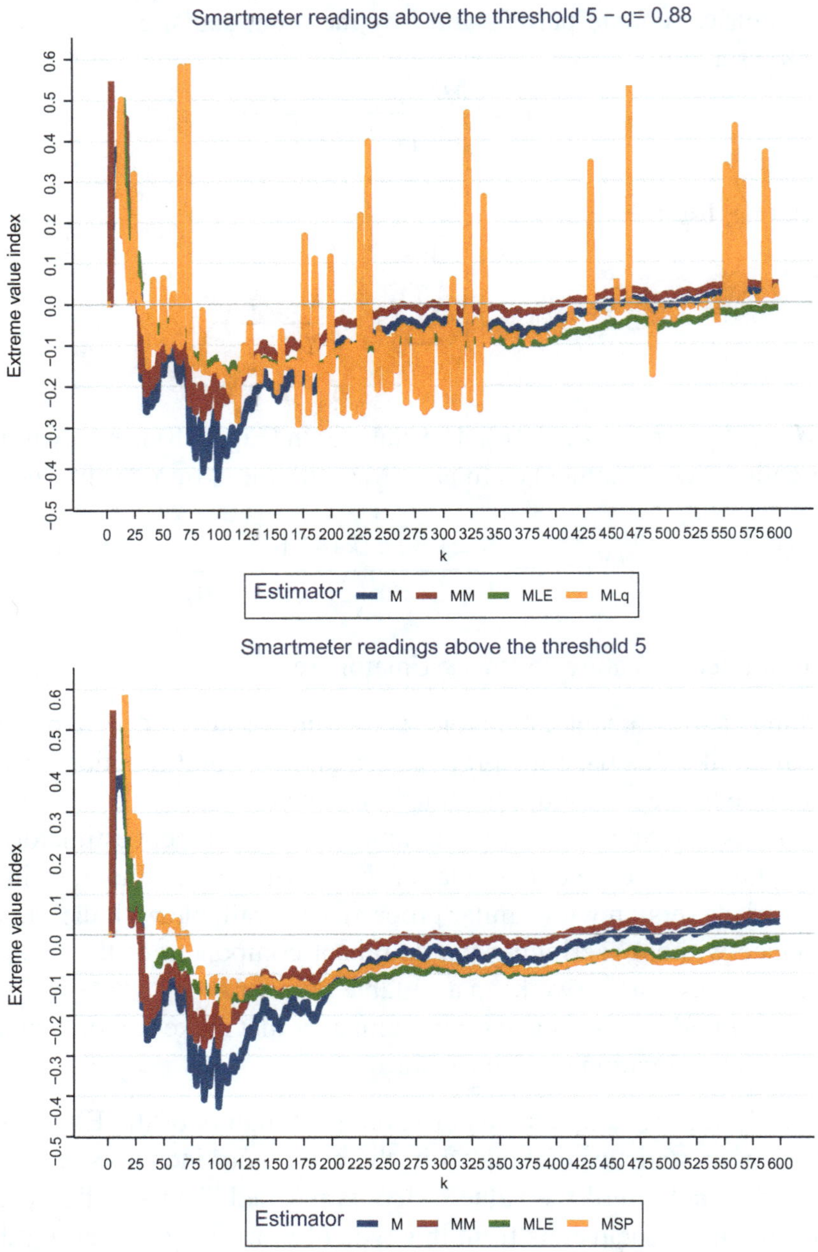

Fig. 4.6 Estimation of the EVI with the POT method

irregular cases which may compromise the practical applicability of ML. Theoretical and numerical accounts of these issues can be found in [37, 38] and references therein.

In the second panel in Fig. 4.6, we swap the MLq estimator with the MPS (or MSP) estimator for the GPD. Although there are issues in the numerical convergence for small values of k, where the variance is stronger, this estimator shows enhanced behavior returning estimates of the EVI in agreement with the remainder estimators. Therefore, it seems reasonable to settle with the point estimate $\hat{\gamma} = -0.01$. It is worth highlighting that the MLq shows its best performance within the corresponding region

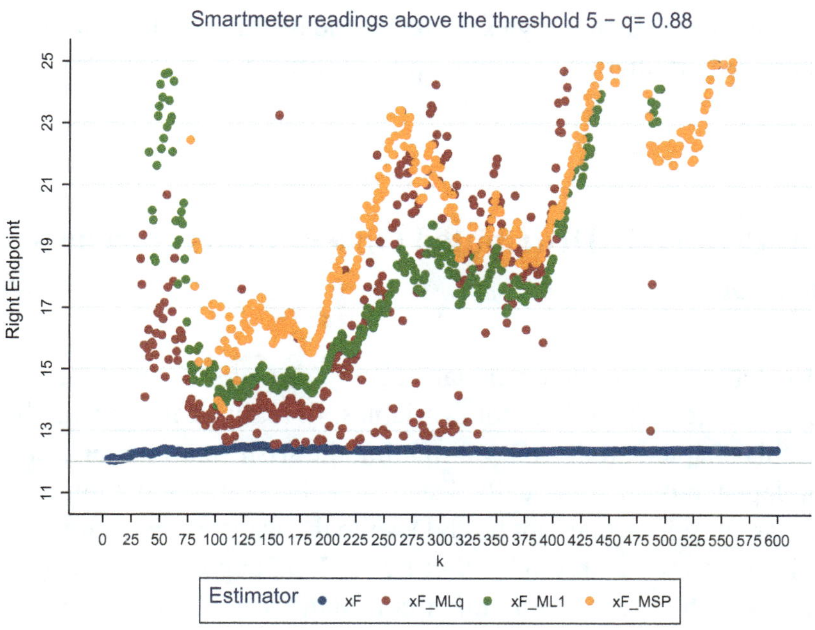

Fig. 4.7 Estimation of the upper endpoint with the POT method

of values of, that is, for k between 125 and 175, a region that also holds feasible for the tests expounded in Sect. 4.3.1.

There is however one estimator for the upper endpoint x^F that does not depend on the estimation of the EVI γ, worked out in [18], this being designed for distributions with finite upper endpoint enclosed in the Gumbel domain of attraction. The so-called general right endpoint estimator is defined as

$$\hat{x}^F := X_{n,n} + X_{n-k,n} - \frac{1}{\log 2} \sum_{i=0}^{k-1} \log\Big(1 + \frac{1}{k+i}\Big) X_{n-k-i,n}. \tag{4.21}$$

Figure 4.7 displays the estimate-yields for several endpoint estimators in the class (Eq. 4.17), with accompanying general endpoint estimator. Again, the corresponding maximum Lq-likelihood estimator is found with the distortion parameter q being set equal to 0.88, where the k-values for which the Lq-likelihood method experienced convergence issues in the estimation of the EVI are now omitted. The value $q = 1$ determines the mainstream ML estimator for the endpoint which the class of estimators defined in Eq. (4.17) also encompasses. The relative finite sample performance of these endpoint estimators is here compared with the naïve maximum estimator $X_{n,n}$. We recall that the observed maximum is 12.01. The general endpoint estimator consistently returns values around 12.4 for almost all values of k. All the other estimators, expect the MLq estimator for the upper endpoint seem to exacerbate the upper bound for the electricity load. Therefore, we find reasonable to advise that the estimate for the upper endpoint of $\hat{x}^F = 13.0$ kWh should be taken as benchmark for future work concerning technologies to enable end use energy demand. We also issue the cautionary remark that this is a mere point estimate, disposed of any confidence bounds.

4.4 Non-identically Distributed Observations—Scedasis Function

Thus far we have assume that our data consists of observations of i.i.d random variables $X_1, \ldots, X_n$ are i.i.d random variables. In reality, this may not be the case. Thus far in this chapter, we have been treating the Irish smart meter data as if these comprise independent and identically distributed observations, but intuitively at least one of these assumptions may not hold. Despite the ways we are sampling extremes from the original raw data, through either BM or POT approaches, inevitably though a few households may end up as main contributors to the extreme values being taken up to the statistical analysis. Figure 4.8 is a representation of the excess load per household above 7 kWh. The top panel shows the exceedances in terms of their cumulative yields over 7 weeks, whilst the bottom panel is the same scatter plot already presented in Fig. 4.2. It is noticeable that the household yielding the absolute maximum of 12.01 kWh does not show the largest total in the cumulative excess plot but two other households are sparking up totals by consistently exceeding the threshold over the course of the 7 weeks. Despite this does not shift the upper endpoint itself (as this is kept fixed by structural settlements with the energy provider), it may push the observations closer to the true upper bound signifying a trend is present in the extreme layers of the process generating the data.

Social demographics and behavioural changes are not likely to occur within the time span considered to this illustrative example, but we can see that in other applications to electricity consumption even a decade is enough time for human behaviour to be vastly different. Regulation of the gas and electricity industry, introduction of software applications that can monitor and incentivise certain consumption over others, time-of-use tariffs, new low carbon technologies and the variety of electronic devices in homes will change the way consumers interact with the grid and consume their electricity and most probably has already changed.

Nonetheless we still want to be able to address the same questions as we did before and we want to ensure that there is a probabilistic framework that grounds our statistical analysis. Our main concern lies with observations that are not identically distributed but we will include a short review for data that exhibit serial dependence.

There are of course many ways to look at dependence in data sets indexed in time or space. Ways of pre-processing data to alleviate dependence issues and possible non-stationary have been considered in [39]. Historically, however, simpler clustering techniques have been employed [40]. We already discussed apropos to the BM, to choose the length of block in such a way that the dependence no longer plays a role or is weak for the extreme value theorem to hold. Ensuring that our observation

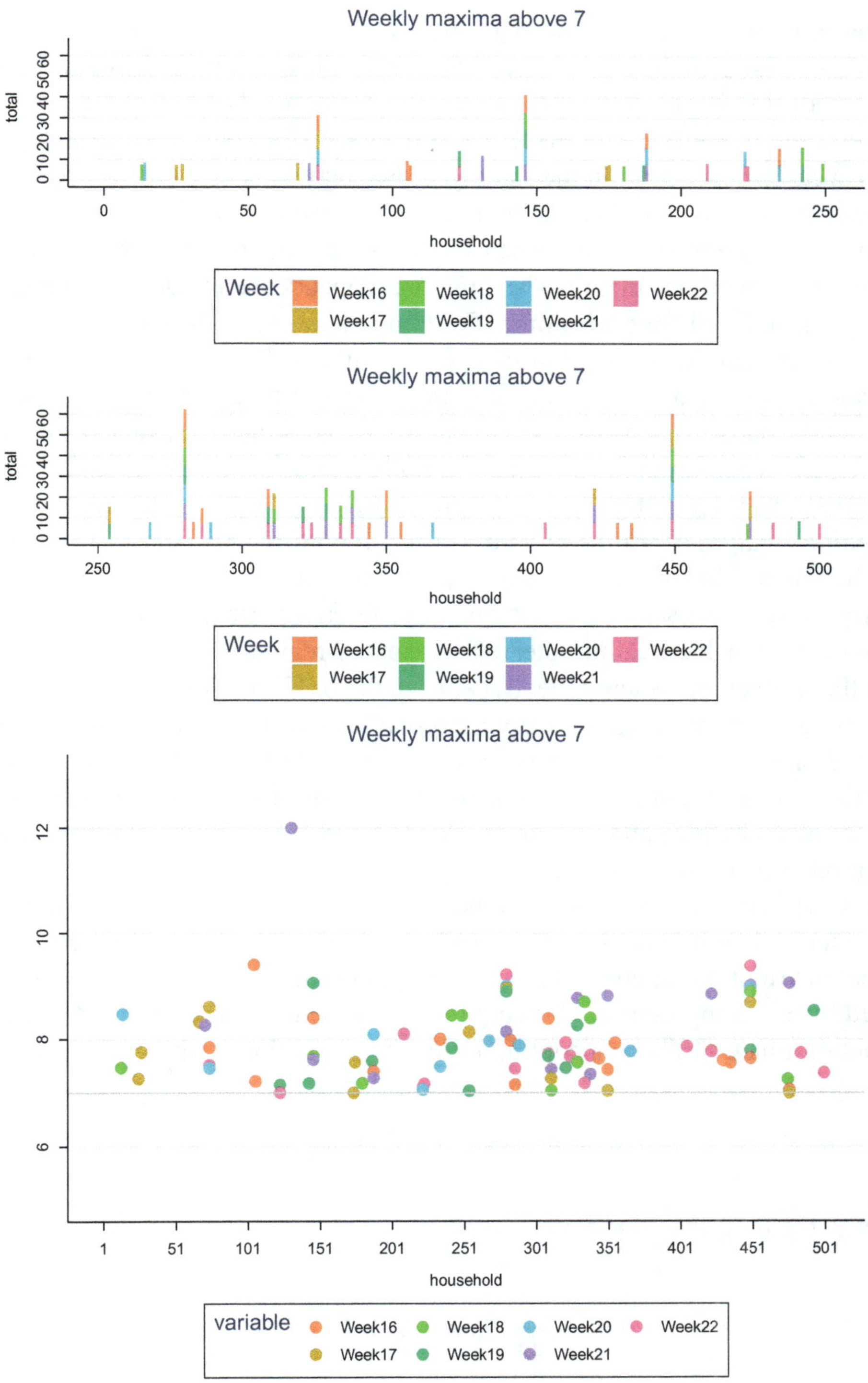

Fig. 4.8 Estimation of the upper endpoint with the POT method

come from separate events is a simple way of ascertaining independence and that the sampled data contains meaningful information on the widely different phenomena emanating from similar hazards (earthquakes, rainfall, storms). For example, if we were considering heatwaves, they may last up to a fortnight thus considering daily maxima in temperature taken during a heatwave will be part of the same weather system and subsequently dependent and also less representative of the wild. Similarly, if we are considering rainfall, a low pressure system may last two or three days thus maxima taken every 2–3 non-overlapping days may be considered to independent or only weakly dependent. In the same vein, we may look at weekly maxima of the electric load profiles of individual household to weed out the daily and sub-weekly patterns. Thus, the block to be chosen should be application and data specific.

However, sometimes there is also temporal dependence in addition to individual events i.e. profiles change with time. For temperature data, the change comes with the season as well due to the diurnal cycle. Similarly, for electricity data there are many sources of seasonality to consider: the impact from temperature i.e., an annual cycle and the daily cycle as well as human behaviour which may repeat weekly. For example, we may restrict to only taking weekly maxima from the summer, etc. This is where we turn focus to non-stationarity extremes, meaning that the underlying distribution changes over time or across space or both. This aspect will be exploited through the definition of a trend in the frequency of extremes in such a way to maintain integrity as we move across the potentially different distribution functions ascribed to each of the considered time-intervals (or spatial-regions), assumed homogeneous within themselves and heterogeneous between them. The basic structural assumption to the trend on the time-space evolving probability that a certain large outcome is exceeded originates from the concept of comparable tails [41]. There have been estimators accounting for the heteroscedastic (non-stationarity in the scale) setting, first introduced by [42] and further developed by [43] to address challenged arising in the modelling of heavy-tails.

The setup is laid out as follows. Suppose that $X_1^{(n)}, \ldots, X_n^{(n)}$ are observations from independent random variables taken at n time points, which are assumed to follow different distribution functions $F_{n,1}, \ldots, F_{n,n}$ but sharing a common upper endpoint denoted by x^F. Suppose the following limit relation involving an offset or baseline distribution function F and a continuous positive function c taking values in $[0, 1]$,

$$\lim_{x \to x^F} \frac{1 - F_{n,i}(x)}{1 - F(x)} = c\Big(\frac{i}{n}\Big), \tag{4.22}$$

subject to the unifying condition

$$\int_0^1 c(s)d(s) = 1.$$

Doing so allows c to represent the frequency of extremes in the tail. Einmahl et al. [43] advocate a kernel density estimator as ideally suited to tackle the estimation of the scedasis function which can be viewed as a density in itself. Specifically, the estimator for $c(s)$ is given by

$$\hat{c}(s) = \frac{1}{kh} \sum_{i=1}^{n} I_{\{X_i^{(n)} > X_{n,n-k}\}} G\Big(\frac{s - \frac{i}{n}}{h}\Big), \tag{4.23}$$

where G is a continuous, symmetric kernel such that $\int_{-1}^{1} G(s)ds = 1$, with $G(s) = 0$ for $|s| > 1$. The bandwidth, $h := h_n$ satisfies $h \to 0$ and $kh \to \infty$ as $n \to \infty$, and $I_A = 1$ denotes indicator function which is equal to 1 if A holds true and equal to 0 otherwise. Finally we note that $X_{n,n-k}$ is the global random threshold determined by the kth largest observation. We shall defer the estimation of the scedasis in practice to Chap. 5 using the Thames Valley Vision data.

References

1. Beirlant, J., Goegebeur, Y., Segers, J., Teugels, J.L.: Statistics of Extremes: Theory and Applications. Wiley (2004)
2. Castillo, E., Hadi, A., Balakrishnan, N., Sarabia, J.M.: Extreme Value and Related Models with Applications in Engineering and Science. Wiley, Hoboken, New Jersey (2005)
3. de Haan, L,. Ferreira, A.: Extreme Value Theory: An Introduction. Springer (2006)
4. Gumbel, E.J.: Statistics of Extremes. Columbia University Press, New York and London (1958b)
5. Melinda Gálfi, V., Bódai, T., Lucarini, V.: Convergence of extreme value statistics in a two-layer quasi-geostrophic atmospheric model. Complexity (2017)
6. Bingham, N., Goldie, C., Teugels, J.: Regular Variation. Cambridge University Press (1987)
7. de Haan, L.: On regular variation and its application to the weak convergence of sample extremes. Ph.D. thesis, Mathematisch Centrum Amsterdam (1970)
8. Ferreira, A., de Haan, L.: On the block maxima method in extreme value theory: PWM estimators. Ann. Statist. **43**(1), 276–298 (2015)
9. Jeffree, C., Neves, C.: Tilting maximum Lq-likelihood estimation for extreme values drawing on block maxima. Technical report (2018). arXiv:1810.03319
10. Ferrari, D., Yang, Y.: Maximum Lq-likelihood estimation. Ann. Statist. **38**(2), 753–783 (2010)
11. Ferrari, D., Paterlini, S.: The maximum Lq-likelihood method: an application to extreme quantile estimation in finance. Methodol. Comput. Appl. Probab. **11**, 3–19 (2009)
12. Cheng, R.C.H., Amin, N.A.K.: Estimating parameters in continuous univariate distributions with a shifted origin. J. Roy. Statist. Soc. Ser. B **45**, 394–403 (1983)
13. Ranneby, B.: The maximum spacing method. An estimation method related to the maximum likelihood method. Scand. J. Statist. **11**, 93–112 (1984)
14. Ekström, M.: Consistency of generalized maximum spacing estimates. Scand. J. Statist. **28**(2), 343–354 (2001)
15. Huang, C., Lin, J.-G.: Modified maximum spacings estimator for generalized extreme value distribution and applications in real data analysis. Metrika **77**, 867–894 (2013)
16. Balkema, A.A., de Haan, L.: Residual life time at great age. Ann. Prob. 792–804 (1974)
17. Pickands, J.: Statistical inference using extreme order statistics. Ann. Stat. 119–131 (1975)
18. Fraga Alves, I., Neves, C.: Estimation of the finite right endpoint in the Gumbel domain. Statistica Sinica **24**, 1811–1835 (2014)
19. Galambos, J.: A statistical test for extreme value distributions. In: Gnedenko, B.V. et al. (eds.) Non-parametric Statistical Inference, North Holland, Amesterdam, pp. 221–230 (1982)
20. Castillo, E., Galambos, J., Sarabia, J.M.: The selection of the domain of attraction of an extreme value distribution from a set of data. In: Hüsler, J., Reiss, R.-D. (eds.) Extreme Value Theory. Lecture Notes in Statistics, vol. 51, pp. 181–190 (1989)
21. Hasofer, A.M., Wang, Z.: A test for extreme value domain of attraction. J. Am. Stat. Assoc. **87**, 171–177 (1992)
22. Fraga Alves, M.I., Gomes, M.I.: Statistical choice of extreme value domains of attraction—a comparative analysis. Commun. Stat. Theory Methods **25**(4), 789–811 (1927)
23. Wang, J.Z., Cooke, P., Li, S.: Determination of domains of attraction based on a sequence of maxima. Austr. J. Stat. **38**(2), 173–181 (1996)

24. Marohn, F.: An adaptive test for Gumbel domain of attraction. Scand. J. Stat. **25**(2), 311–324 (1998a)
25. Marohn, F.: Testing the gumbel hypothesis via the pot-method. Extremes **1**(2), 191–213 (1998b)
26. Segers, J., Teugels, J.: Testing the gumbel hypothesis by Galton's ratio. Extremes **3**(3), 291–303 (2000)
27. Neves, C., Picek, J., Alves, M.F.: The contribution of the maximum to the sum of excesses for testing max-domains of attraction. J. Stat. Plan. Inference **136**(4), 1281–1301 (2006)
28. Neves, C., Fraga Alves, M.I.: Semi-parametric approach to the hasofer-wang and greenwood statistics in extremes. Test **16**(2), 297–313 (2007)
29. Fraga Alves, I., Neves, C., Rosário, P.: A general estimator for the right endpoint with an application to supercentenarian women's records. Extremes **20**(1), 199–237 (2017)
30. Neves, C., Pereira, A.: Detecting finiteness in the right endpoint of light-tailed distributions. Stat. Probab. Lett. **80**(5), 437–444 (2010)
31. Smith, R.L.: Estimating tails of probability distributions. Ann. Stat. **15**, 1174–1207 (1987)
32. Dekkers, A.L.M., Einmahl, J.H.J., de Haan, L.: A moment estimator for the index of an extreme-value distribution. Ann. Stat. **17**, 1833–1855 (1989)
33. Fraga Alves, M.I., Gomes, M.I., de Haan, L., Neves, C.: Mixed moment estimator and location invariant alternatives. Extremes **12**(2), 149–185 (2009)
34. Drees, H., Ferreira, A., de Haan, L.: On maximum likelihood estimation of the extreme value index. Ann. Appl. Probab. **14**(3), 1179–1201 (2004)
35. Qi, Y., Peng, L.: Maximum likelihood estimation of extreme value index for irregular cases. J. Stat. Plan. Inference **139**, 3361–3376 (2009)
36. Zhou, C.: The extent of the maximum likelihood estimator for the extreme value index. J. Multivar. Anal. **101**(4), 971–983 (2010)
37. Castillo, J., Daoudi, J.: Estimation of generalized Pareto distribution. Stat. Probabiliy Lett. **79**, 684–688 (2009)
38. Castillo, J., Serra, I.: Likelihood inference for generalized Pareto distribution. Comput. Stat. Data Anal. **83**, 116–128 (2015)
39. Eastoe, E.F., Tawn, J.A.: Modelling non-stationary extremes with application to surface level ozone. J. R. Stat. Soc.: Ser. C (Appl. Stat.) **58**(1), 25–45 (2009)
40. Coles, S., Bawa, J., Trenner, L., Dorazio, P.: An Introduction to Statistical Modeling of Extreme Values, vol. 208. Springer (2001)
41. Resnick, S.I.: Tail equivalence and its applications. J. Appl. Prob. **8**, 135–156 (1971)
42. de Haan, L., Tank, A.K., Neves, C.: On tail trend detection: modeling relative risk. Extremes **18**(2), 141–178 (2015)
43. Einmahl, J.H., Haan, L., Zhou, C.: Statistics of heteroscedastic extremes. J. R. Stat. Soc.: Ser. B (Stat. Methodol.) **78**(1), 31–51 (2016)

Chapter 5
Case Study

5.1 Predicting Electricity Peaks on a Low Voltage Network

In the previous chapter, we looked at load measurements for all households together and we ignored their chronological order. In contrast, in this chapter, we are interested in short term forecasting of household profiles individually. Therefore, information about the time at which measurements were taken becomes relevant.

To illustrate different popular methods and look at their errors, we use a subset of the End point monitoring of household electricity demand data from the Thames Valley Vision project,[1] kindly made publicly available by Scottish and Southern Energy Networks,[2] containing profiles for 226 households in 30 min resolution.

We use the time window from Sunday, 19 July 2015 00.00 to Monday, 21 September 2015 00.00. The first eight weeks (or less depending on the model) are used for training the models, and we want to predict the ninth week (commencing on the 14th September 2015), each half hour usage in that week for each of the households.

On the Fig. 5.1, mean value (top) or maximum value (bottom) for each half-hour is computed for the day of the week for each household, and a box-plot over households is presented. On the Fig. 5.2, a box-plot is produced for each household over all the values recorded during the eight weeks of observations for that household.

The data for each household consists of 3072 half-hourly values. We want to predict the last 336 value in each time series. The exploratory data analysisconfirms that there is a daily seasonality detected. The examples of seasonal decomposition using an additive seasonal model with a lag of 48 half-hours, i.e. one day, and auto-correlation and partial auto-correlation functions with lag 48 for two households are given on Fig. 5.3. In most cases, no uniform trend is observed, daily seasonal component usually contains morning and evening peak as expected, and residuals

[1] https://www.thamesvalleyvision.co.uk.

[2] http://data.ukedc.rl.ac.uk/simplebrowse/edc/Electricity/NTVV/EPM.

M. Jacob et al., *Forecasting and Assessing Risk of Individual Electricity Peaks*, Mathematics of Planet Earth,
https://doi.org/10.1007/978-3-030-28669-9_5

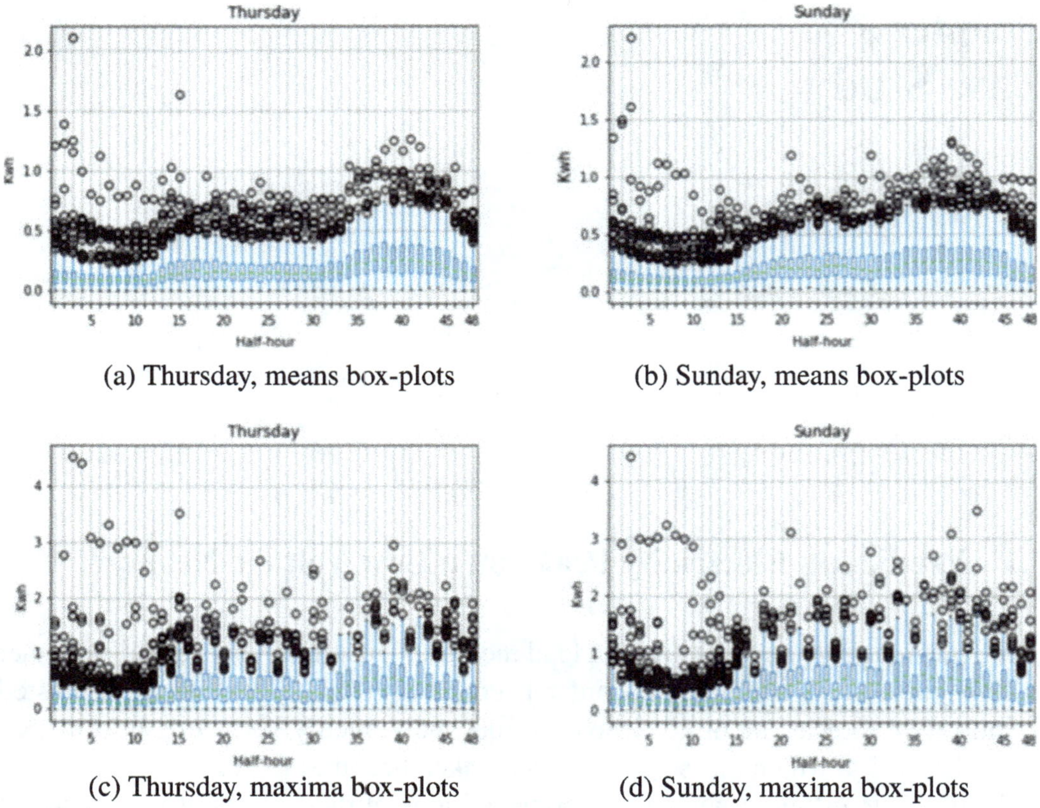

(a) Thursday, means box-plots

(b) Sunday, means box-plots

(c) Thursday, maxima box-plots

(d) Sunday, maxima box-plots

Fig. 5.1 Usage on different days

mostly look random, as expected. The autocorrelation and partial autocorrelation inform us on how many past observations are relevant for predicting a half-hour, and they look different for different households.

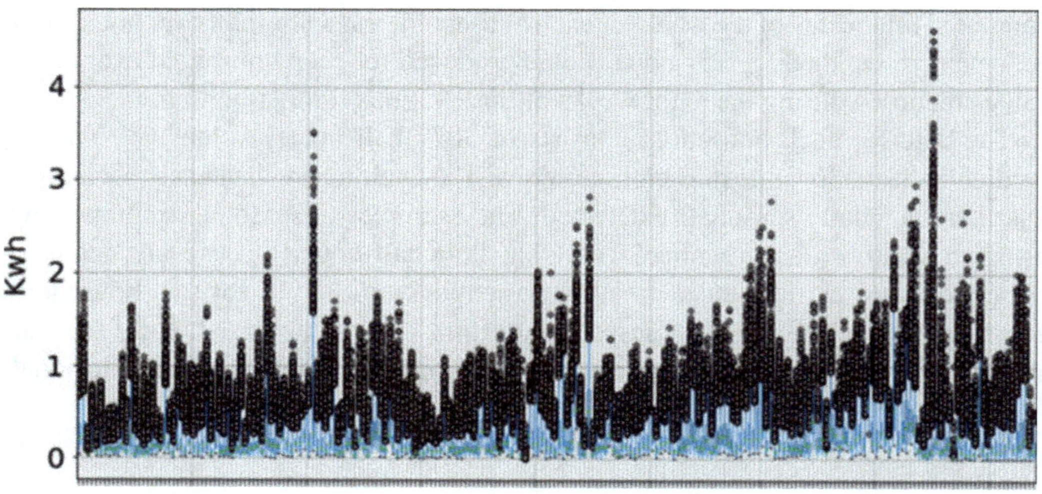

Fig. 5.2 Households-half-hourly usage box-plot

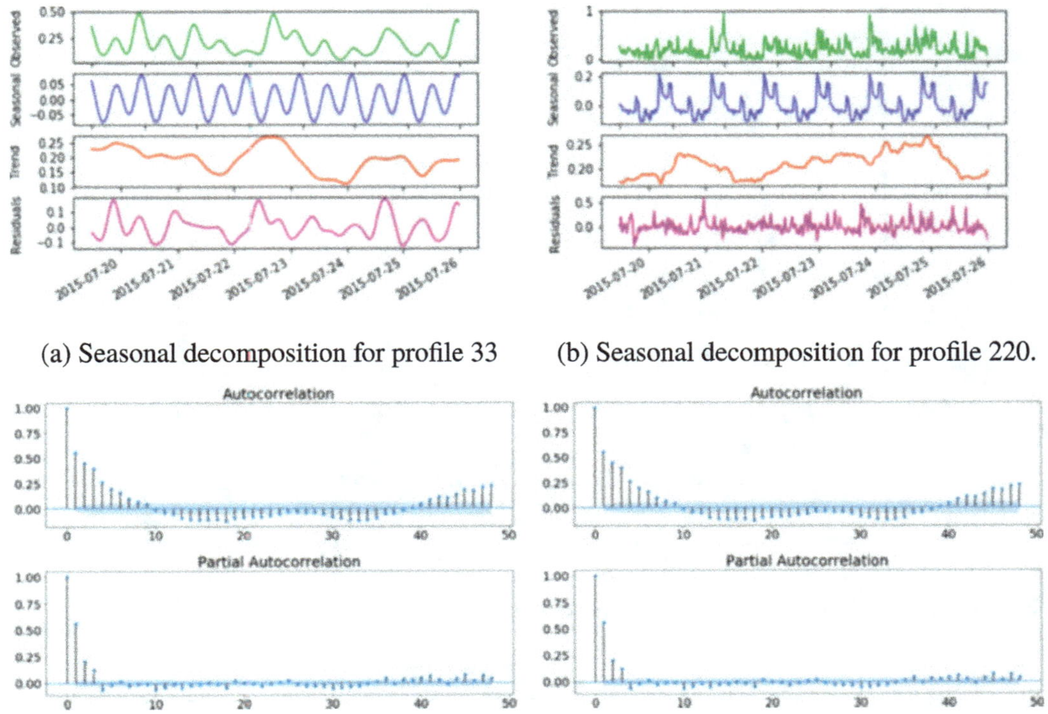

(a) Seasonal decomposition for profile 33

(b) Seasonal decomposition for profile 220.

(c) ACF and partial ACF for household 33.

(d) ACF and partial ACF for household 220.

Fig. 5.3 Exploratory data analysis of household profiles

5.1.1 Short Term Load Forecasts

In this section, several different popular forecasting algorithms from both statistical and machine learning backgrounds will be tested. We will evaluate them using four error measures described in Sect. 2.2, MAPE, MAE, MAD and E_4.

Since we want to compare errors for different forecasting algorithms, in Chap. 2 we have established two simple benchmarks. A **last week (LW)** forecast, where the data from one week before is used to predict the same half-hour of the test week is extremely simple (as no calculation is needed), but relatively competitive. A simple average over several past weeks is also popular, a so called **similar day (SD)** forecast.

Deciding on how many weeks of history to use to average over is not always straightforward, especially when seasons change. Here we have done a quick optimisation of the number of weeks used. Although the smallest error is obtained for one week, i.e. when SD is the same to LW forecast, we use four weeks of history, as this resulted in the smallest 4th norm error, and we are interested in peaks. Examples of LW and SD forecasts are given on Figs. 5.4 and 5.5.

In addition to the two benchmarks, LW and SD, four different algorithms: SARIMA (seasonal ARIMA), Permutation merge (PM), LSTM (recurrent neural network) and MLP (forward neural network) are compared. The detailed descriptions of the algorithms are given in Chap. 2.

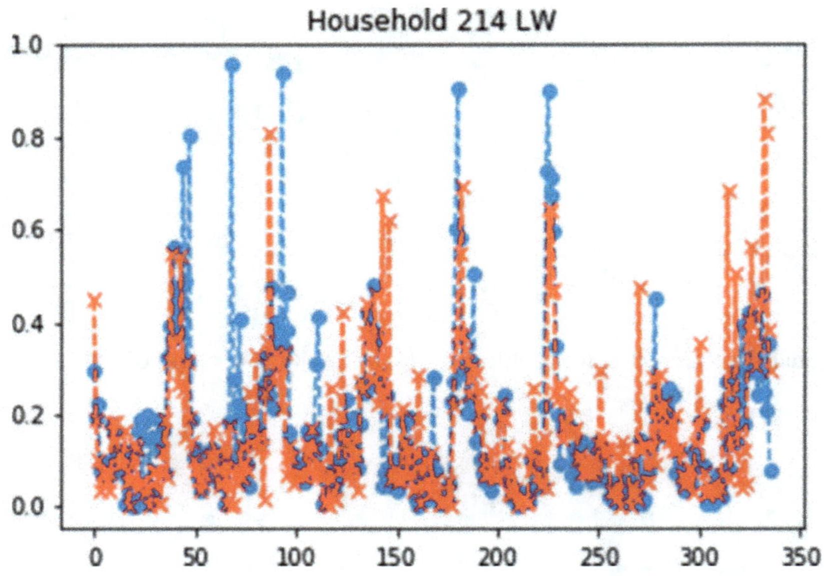

Fig. 5.4 The LW forecast (red cros) and observation (blue dot) for one household

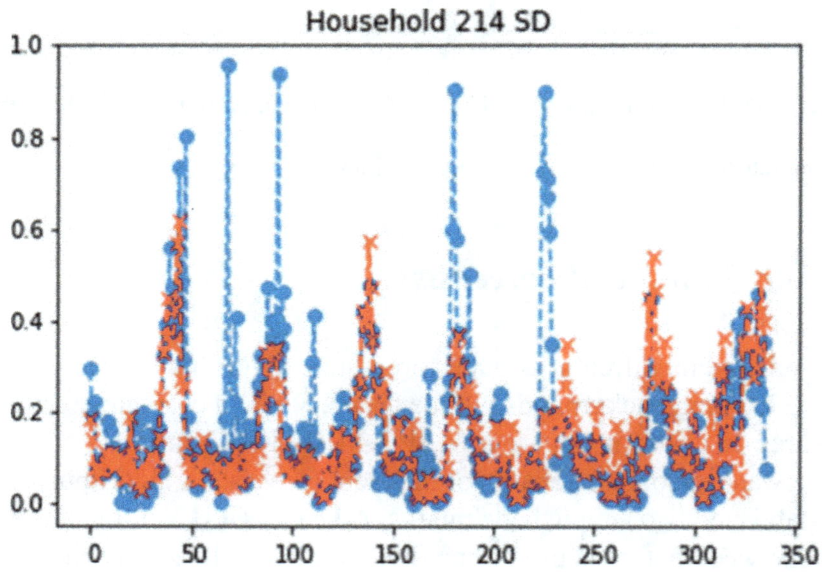

Fig. 5.5 The SD forecast (red cros) and observation (blue dot) for one household

5.1.1.1 SARIMA

As previously discussed, this is a variation of a widely used ARIMA model, where the past values are used to predict future, but also moving average helps to pick up changes in the observations. Integration ensures stationarity of the data. In seasonal autoregressive integrated moving average model (SARIMA) , seasonal part is added. In our case, that is the detected daily seasonality. The time series is split into peak load and seasonal part. The general peak load is assumed to be without and seasonal part is with periodicity. The parameters we use are $p = \{2, 3, 5, 6\}$, $d = 1$, $q = \{0, 1, 3\}$

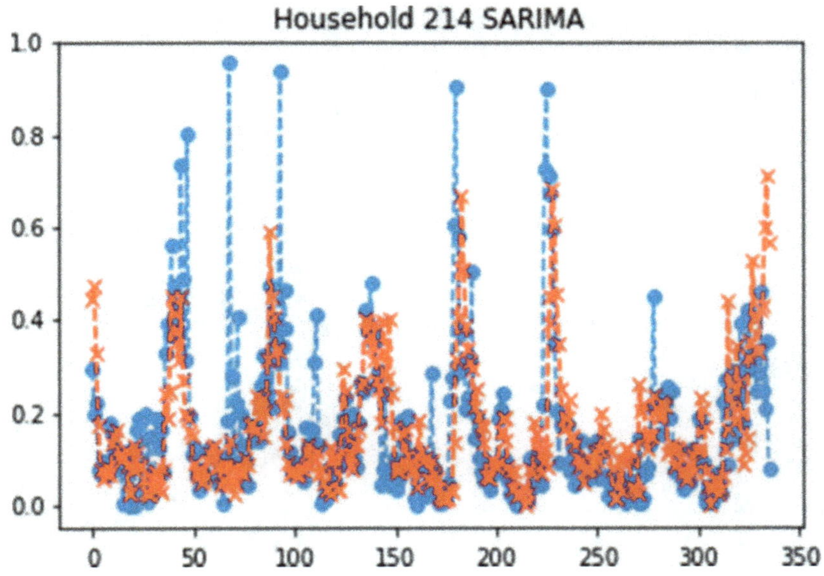

Fig. 5.6 The SARIMA forecast (red cros) and observation (blue dot) for one household

for the general part and $P = \{1, 2\}$, $D = 0$, $Q = \{0, 1, 2\}$ for the seasonal part of the model. The parameters were obtained doing localised search based on the Akaike Information Criterion (AIC) for each household.[3] An example showing some success with the prediction of peaks can be seen on Fig. 5.6.

5.1.1.2 PM

The algorithm with the size of the window 1, therefore allowing permutations with one half hour before and after, was run for a different number of weeks in history. When using only one week, this is equal to LW benchmark. As shown on Fig. 5.7, there was no single value that optimised all four errors. We have chosen 4 weeks of history based on the smallest E_4 error. While relatively similar to SD forecast, PM manages to capture some peak timings better (compare Fig. 5.5 with Fig. 5.8).

5.1.1.3 LSTM

The Long Short Term Memory, a recurrent neural network method with two hidden layers with 20 nodes each was used, implemented with Python Keras library [1], training the model over 5 epochs and using a batch size of 48 (the number of samples per gradient update). The optimiser used was 'adam', a method for first-order gradient-based optimization of stochastic objective functions, based on adaptive estimates of lower-order moments [2]. For the input, we have used the previous load, half-hour and day of the week values. This was coded by 4 values: 0 for working

[3]We used Pmdarima Python library, and its auto_arima functionality.

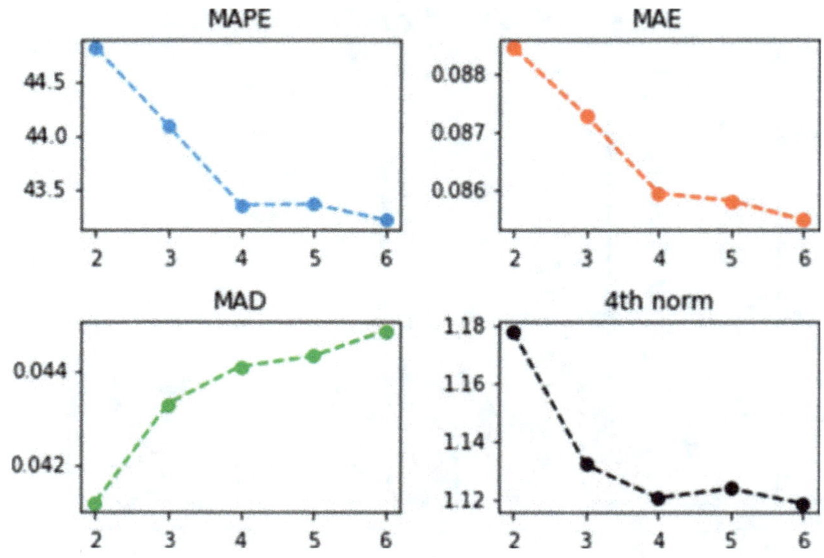

Fig. 5.7 PM algorithms performance for different mean error values

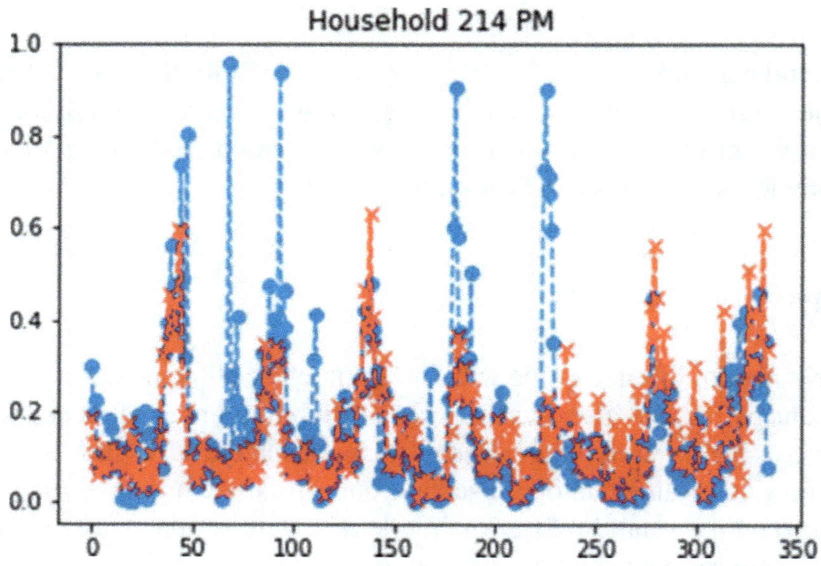

Fig. 5.8 The PM4 forecast (red cros) and observation (blue dot) for one household

day followed by working day; 1 for working day followed by non-working day; 2 for non-working day followed by working day; 3 for non-working day followed by non-working day.

We ran limited parameter search from 10 to 30 nodes in each layer, and noticed, similar to [3] that the equal number of nodes per layer seem to work best. The minimal errors for all four error measures were obtained for the configuration with 20 nodes in each hidden layer, which agrees with the optimal choice obtained by [4], based on MAPE error only. An example of LSTM forecast can be seen on Fig. 5.9.

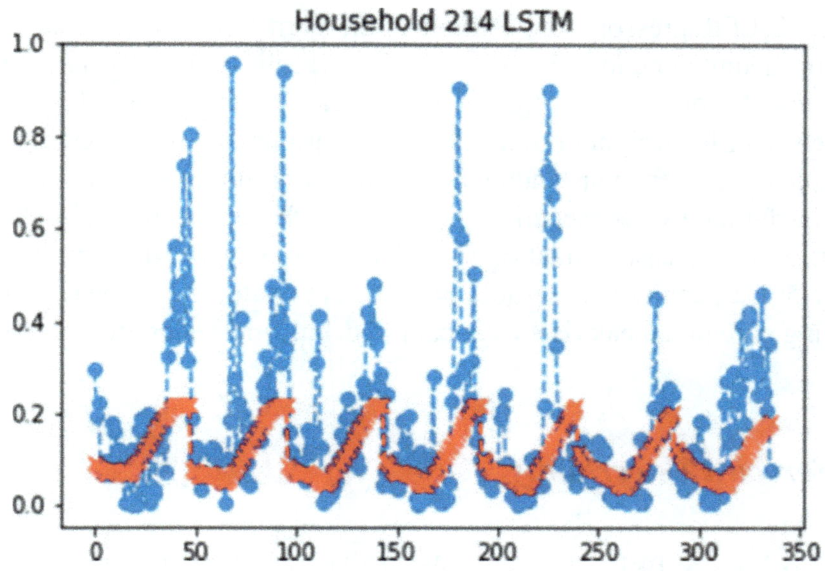

Fig. 5.9 The LSTM forecast (red cros) and observation (blue dot) for one household

5.1.1.4 MLP

Multi-layer perceptron, a feed-forward neural network with five hidden layers and nine nodes in each was chosen, after running limited parameter search, firstly deciding on the number of nodes in two layers (from 5 to 20) and then adding layers with the optimal number of neurons, 9, until the errors started to grow. All four error measures were behaving the same way.

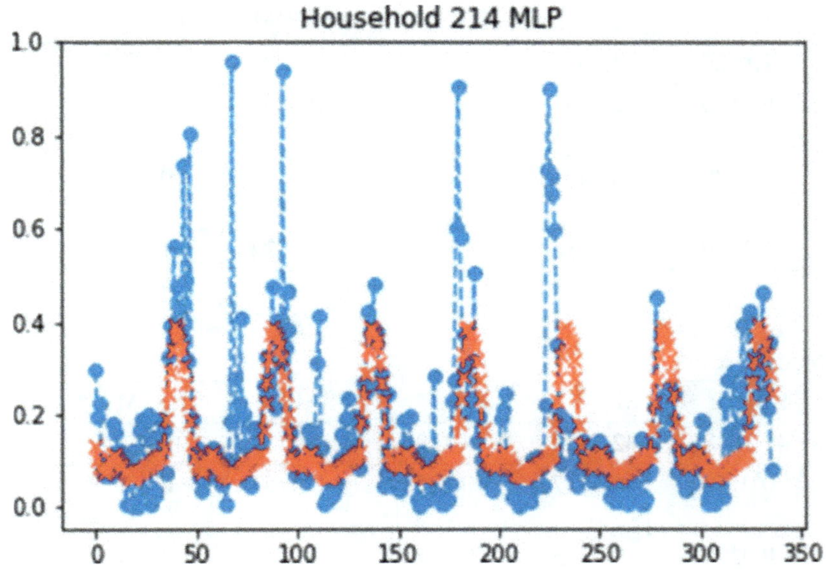

Fig. 5.10 The MLP forecast (red cros) and observation (blue dot) for one household

We used MLPRegressor from Python scikit-learn library [5], using 'relu', the rectified linear unit function, $f(x) = \max(0, x)$ as the activation function for the hidden layers. An optimiser in the family of quasi-Newton methods, 'lbfgs', was used as the solver for weight optimisation. The learning rate for weight updates was set to 'adaptive', i.e. kept constant on 0.01, as long as training loss was continuing to decrease. Each time two consecutive epochs fail to decrease training loss by at least 0.0001, or fail to increase validation score by at least 0.0001, the learning rate was divided by 5. L2 penalty was set to $\alpha = 0.01$. An example is shown on Fig. 5.10, where timing of regular peaks is mostly captured, but amplitudes are underestimated.

5.1.2 *Forecast Uncertainty*

In Tables 5.1, 5.2, 5.3, respectively, means, medians and maxima over households of all four errors are given for the four algorithms and two benchmarks. The box-plot of four errors means across 226 households is given on Fig. 5.11. The results show that SARIMA forecast is having the smallest errors for E_4 error measure and performing best with respect to peaks. Two benchmarks are very competitive, when looking across all the values, with LW doing very well in all other three error measures. PM and MLP are slightly worse and LSTM is lagging behind.

While four error measures give values that are all positive, the differences between predicted and actual value can be negative, in the case of underestimation. This is of importance, especially for peaks. The consequences of underestimated peaks

Table 5.1 Mean errors

Error	LW	SD	PM	SARIMA	LSTM	MLP
MAPE (%)	**41.8924**	44.3813	45.1576	43.3889	47.1202	47.7836
MAE (kWh)	**0.0836**	0.0844	0.0861	0.0866	0.0957	0.0902
MAD (kWh)	**0.0329**	0.04314	0.0440	0.0450	0.0512	0.0538
E_4 (kWh)	1.2299	1.0588	1.0774	**1.0524**	1.2307	1.0924

Table 5.2 Median errors

Error	LW	SD	PM	SARIMA	LSTM	MLP
MAPE (%)	42.7588	**42.3669**	43.1196	44.0777	45.2884	43.8456
MAE (kWh)	0.0742	**0.0737**	0.0758	0.0757	0.0792	0.0754
MAD (kWh)	**0.0278**	0.0363	0.0375	0.0387	0.0445	0.0454
E_4 (kWh)	1.1785	**0.9792**	0.9916	1.0127	1.1598	1.0183

Table 5.3 Maximum errors

Error	LW	SD	PM	SARIMA	LSTM	MLP
MAPE (%)	150.3073	463.8288	459.1680	**145.5191**	343.5002	681.7654
MAE (kWh)	**0.2688**	0.4831	0.4823	0.3111	0.7156	0.5721
MAD (kWh)	**0.1365**	0.1680	0.1664	0.1700	0.1742	0.1799
E_4 (kWh)	3.6269	5.7239	5.8357	**3.0036**	8.2499	6.4193

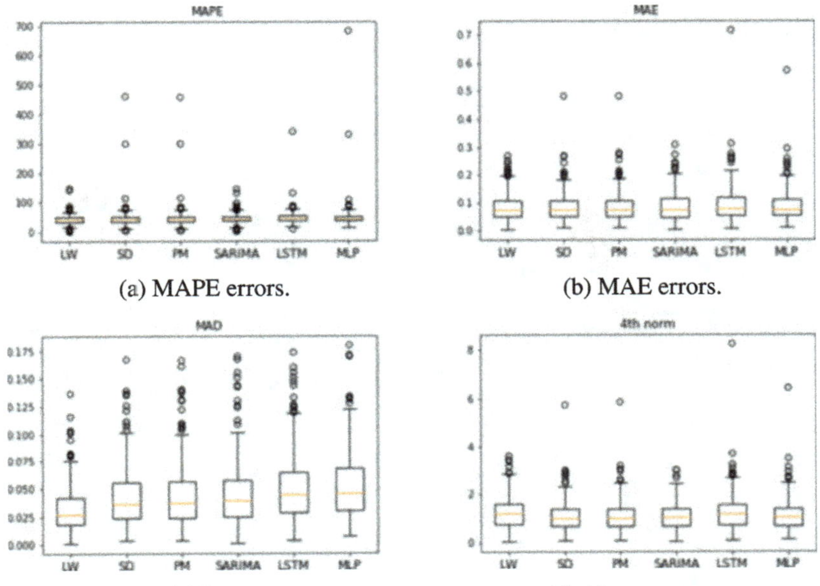

(a) MAPE errors. (b) MAE errors. (c) MAD errors. (d) 4th norm errors.

Fig. 5.11 Average errors across households

(higher prices, outages, or storage control problems) are usually much worse than overestimation of peaks (higher costs, or non-optimal running). Histograms of those distances for all used methods are given on Fig. 5.12 with normal probability distribution function contours, based on the distances' mean and standard deviation. One can see that these distances are not normally distributed. Almost all forecasts are more one-sided, therefore underestimating. This is especially pronounced for SD, PM, LSTM and MLP forecasts. Also, one can notice similarity in distances profiles between LW, and SARIMA on one hand and between other four forecast on the other hand.

We note that this predictive task is quite challenging. In the week commencing 31 Aug 2015, there is a double challenge of a summer bank holiday (31 Aug) and beginning of school year, while summer weeks before that in the UK in general are characterised by less consumption. This leaves only one full week of behaviour relatively similar to the week that we want to predict which explains why LW's MAPE is on average better than more sophisticated methods.

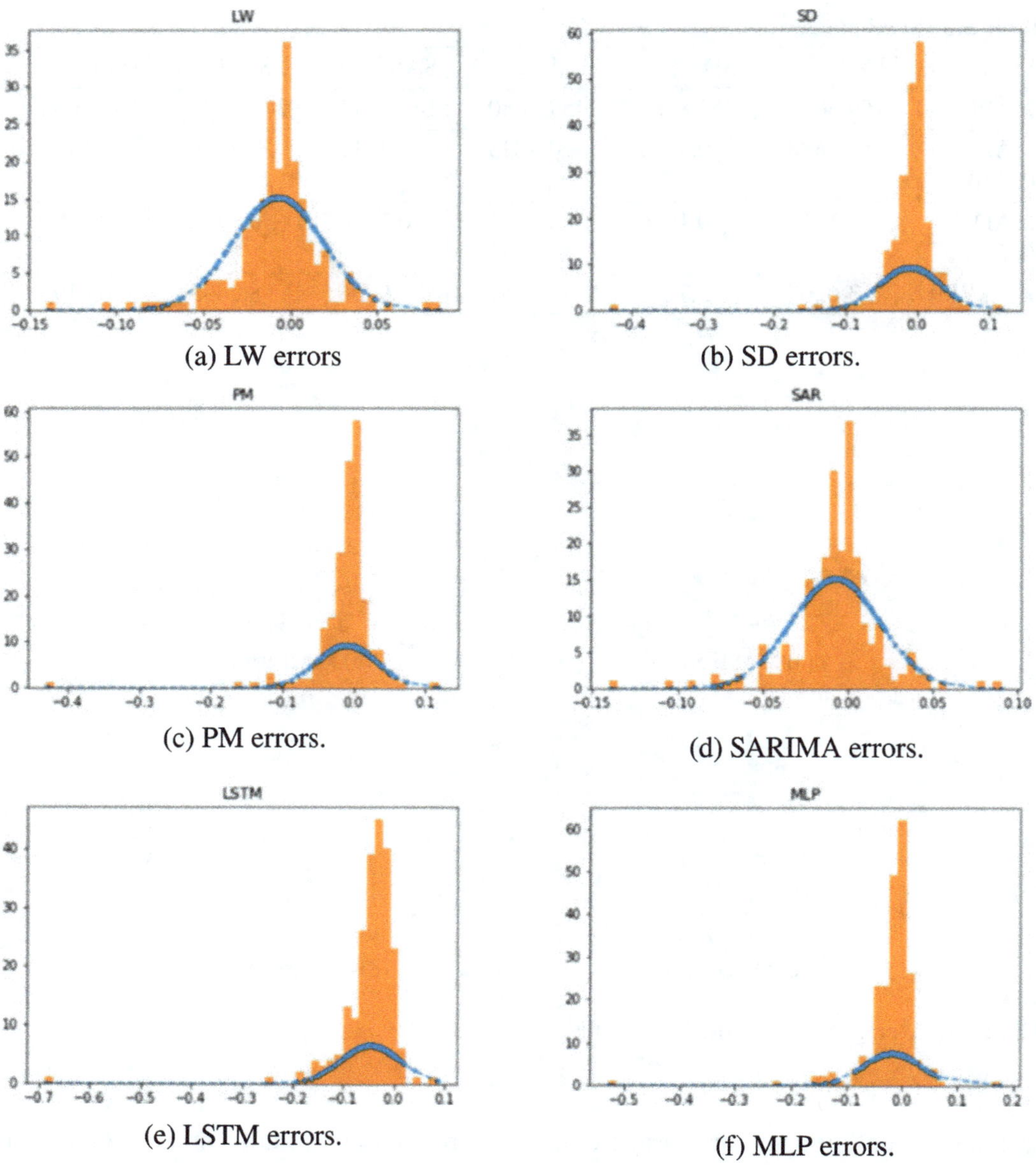

(a) LW errors (b) SD errors. (c) PM errors. (d) SARIMA errors. (e) LSTM errors. (f) MLP errors.

Fig. 5.12 Histogram of errors—differences between predicted and observed values

5.1.3 Heteroscedasticity in Forecasts

In this section, we want to look into timing and frequency of largest errors for all forecast methods that were compared in the previous section. We want to see if we can spot any patterns. Are the different methods better in different time-steps? Can we identify time periods that are more difficult for forecasting? To this end, we apply the development of the scedasis introduced in Sect. 4.4 to capture how the largest absolute errors stemming from each forecasting method evolve over time. The interpretation is the following: the higher the scedasis at time $t \in [0, 1]$, the higher the propensity for extreme errors to occur at that time t. A value around 1 means stationarity in the errors of forecasts.

Figure 5.13 displays the estimated scedasis, as given in (4.22), when we select the largest 50 errors determined by each forecasting method. The SARIMA model yield the least oscillation around 1 which is indicative of satisfactory performance of this time series model in capturing the relevant traits in the data. Both PM4 and SD4 seem to have better predictive power on later days of the week as they exhibit a decreasing trend in the likelihood of large errors. All methods show large uncertainty in the forecasts delivered between Wednesday and Thursday, where all the sample paths for the scedasis tend to concentrate above 1. The early hours of Thursday maxima box-plots on Fig. 5.1c when compared with Fig. 5.1a show more spread in values. In this way, the estimated scedasis values give us a way to quantify which times are more difficult for prediction regarding different algorithms.

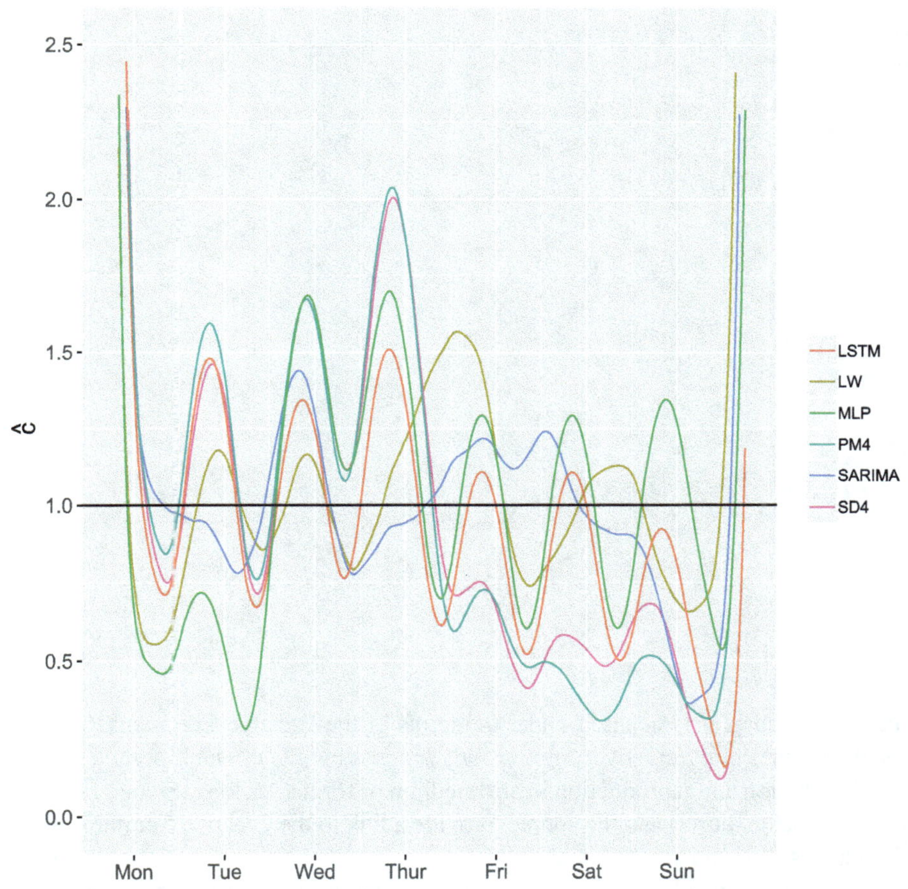

Fig. 5.13 Estimated scedasis, $\hat{c}$, as a function of time

References

1. Chollet, F. et al.: Keras. https://keras.io (2015)
2. Kingma, D., Lei Ba, J.: Adam: a method for stochastic optimization. In: Proceedings of the 3rd International Conference on Learning Representations, ICLR 2015 (2015)
3. Cavallo, J., Marinescu, A., Dusparic, I., Clarke, S.: Evaluation of forecasting methods for very small-scale networks. In: Woon, W.L., Aung, Z., Madnick, S. (eds.) Data Analytics for Renewable Energy Integration, pp. 56–75. Springer International Publishing, Cham (2015)
4. Kong, W., Dong, Z.Y., Jia, Y., Hill, D.J., Xu, Y., Zhang, Y.: Short-term residential load forecasting based on lstm recurrent neural network. IEEE Trans. Smart Grid (2017)
5. Pedregosa, F., Varoquaux, G., Gramfort, A., Michel, V., Thirion, B., Grisel, O., Blondel, M., Prettenhofer, P., Weiss, R., Dubourg, V., Vanderplas, J., Passos, A., Cournapeau, D., Brucher, M., Perrot, M., Duchesnay, E.: Scikit-learn: machine learning in python. J. Mach. Learn. Res. **12**, 2825–2830 (2011)

Index

M. Jacob et al., *Forecasting and Assessing Risk of Individual Electricity Peaks*, Mathematics of Planet Earth,
https://doi.org/10.1007/978-3-030-28669-9

www.ingramcontent.com/pod-product-compliance
Lightning Source LLC
LaVergne TN
LVHW082239200926
848484LV00008B/2378